Editor-in-Chief

HARRY R. ALLCOCK

Department of Chemistry
Pennsylvania State University

●●

INORGANIC SYNTHESES

Volume 25

WILEY

A Wiley-Interscience Publication
JOHN WILEY & SONS

New York   Chichester   Brisbane   Toronto   Singapore

Published by John Wiley & Sons, Inc.

Library of Congress Catalog Number: 39-23015
ISBN 0-471-61874-8

Printed in the United States of America

10 9 8 7 6 5 4 3 2 1

This volume is dedicated to W. Conard Fernelius (1905–1986), outstanding teacher, researcher, and scholar, who was one of the founders of *Inorganic Syntheses* and who continued throughout his life to be a leader in its development.

PREFACE

The synthesis of chemical compounds is the bedrock on which all other areas of chemistry depend. In this sense, *Inorganic Syntheses* represents an evolving compilation of techniques and ideas that provide a cross section of activity in the field at a particular time. Providing such a cross section is no easy task. At the present time, the field of inorganic synthesis is undergoing significant changes. Starting from its traditional foundation in small-molecule nonmetal chemistry and metal coordination chemistry, it has in recent years absorbed the dramatic expansion of activity in transition metal organometallic chemistry and cluster chemistry. More recently, connections have been developed to polymer chemistry, ceramic science, electroactive solids, and pharmacology.

The increasing involvement of inorganic chemists with the fields of polymer chemistry, solid state science, and ceramic science is based on the precept that the experimental techniques, theories, and models of small-molecule inorganic chemistry can be applied to the infinitely more complex molecules found in these other areas. Synthetic chemists can play a major role in the design and preparation of new materials if a significant conceptual hurdle can be overcome. This hurdle is the dividing line between the well-understood behavior of small molecules (2–500 atoms) and the more complex behavior of linear macromolecules and three-dimensional, covalently bonded solids. It is a dividing line between compounds that can be characterized easily by modern methods and those that cannot, and between species with absolute compositions and precise molecular weights and substances with variable formulas and structures that can be understood only in statistical rather than absolute terms. Yet the knowledge and experience available within the community of inorganic chemists could have a profound influence on the development of new supramolecular systems, on which much of our advanced technology increasingly depends.

For this reason I have included in this volume a few examples of inorganic polymer syntheses (Chapter 2) and preparations of inorganic ring

Previous volumes of *Inorganic Syntheses* are available. Volumes I–XVI can be ordered from R. E. Krieger Publishing Co., Inc., P.O. Box 9542, Melbourne, Florida 32901; Volume XVII is available from McGraw-Hill, Inc.; subsequent volumes can be obtained from John Wiley & Sons, Inc.

systems (Chapter 1), which constitute models for more complex molecules and may eventually serve as reaction intermediates for polymers or for main group solid state syntheses. Chapter 3 provides a brief glimpse of the emerging role being played by synthesis in the development of pharmacologically active inorganic compounds. Chapter 4 contains examples of small-molecule coordination complexes and related compounds—the traditional core of inorganic chemistry. In addition, and in accordance with the format of previous volumes of *Inorganic Syntheses,* a section (Chapter 5) is set aside for transition metal organometallic compounds, including species with metal–metal bonds and cluster molecules. In this respect, I have been fortunate to be assisted by Gregory L. Geoffroy, who gathered together and edited the syntheses in this section. I thank him for his contribution to this volume.

Synthetic chemistry of any kind is not a trivial activity, and it should be pursued only by those who can recognize and avoid the inherent safety risks that exist. Thus, most of the procedures given in this volume should be undertaken only by individuals who are already competent synthetic chemists or who are working directly under the close supervision of someone who is. Potential hazards are identified throughout this volume, and these warnings should be taken seriously.

The main purpose of *Inorganic Syntheses* is to provide illustrative synthetic methods that are *reliable.* Thus, all the syntheses reported here have been checked experimentally by independent investigators. I greatly appreciate the contributions of those who submitted the syntheses and the truly essential efforts of the individuals who did the checking. Major assistance with the routing of correspondence and manuscripts to authors and checkers, and maintaining the overall momentum and organization of this volume, was provided by Noreen Allcock. I would also like to thank the members of the Editorial Board for their helpful suggestions and support. The contributions by Thomas Sloan, who compiled the index and answered nomenclature questions, and the advice of Duward Shriver on procedural matters, are particularly appreciated.

HARRY R. ALLCOCK

University Park, Pennsylvania
February 1988

NOTICE TO CONTRIBUTORS
AND CHECKERS

The *Inorganic Syntheses* series is published to provide all users of inorganic substances with detailed and foolproof procedures for the preparation of important and timely compounds. Thus the series is the concern of the entire scientific community. The Editorial Board hopes that all chemists will share in the responsibility of producing *Inorganic Syntheses* by offering their advice and assistance in both the formulation of and the laboratory evaluation of outstanding syntheses. Help of this kind will be invaluable in achieving excellence and pertinence to current scientific interests.

There is no rigid definition of what constitutes a suitable synthesis. The major criterion by which syntheses are judged is the potential value to the scientific community. For example, starting materials or intermediates that are useful for synthetic chemistry are appropriate. The synthesis also should represent the best available procedure, and new or improved syntheses are particularly appropriate. Syntheses of compounds that are available commercially at reasonable prices are not acceptable. We do not encourage the submission of compounds that are unreasonably hazardous, and in this connection, less dangerous anions generally should be employed in place of perchlorate.

The Editorial Board lists the following criteria of content for submitted manuscripts. Style should conform with that of previous volumes of *Inorganic Syntheses*. The introductory section should include a concise and critical summary of the available procedures for synthesis of the product in question. It should also include an estimate of the time required for the synthesis, an indication of the importance and utility of the product, and an admonition if any potential hazards are associated with the procedure. The Procedure should present detailed and unambiguous laboratory directions and be written so that it anticipates possible mistakes and misunderstandings on the part of the person who attempts to duplicate the procedure. Any unusual equipment or procedure should be clearly described. Line drawings should be included when they can be helpful. All safety measures should be stated clearly. Sources of unusual starting materials must be given, and, if possible, minimal standards of purity of reagents and solvents should be stated. The scale should be reasonable for normal laboratory operation, and any problems involved in scaling the procedure either up or down should be discussed. The criteria for judging the purity of the final product should be delineated clearly. The section on Properties should supply and discuss those physical and chemical characteristics that are relevant to judging the purity of the product and to permitting its handling and use in an intelligent manner. Under References, all pertinent literature citations should be listed in order. A style sheet is available from the Secretary of the Editorial Board.

The Editorial Board determines whether submitted syntheses meet the general specifications outlined above, and the Editor-in-Chief sends the manuscript to an independent laboratory where the procedure must be satisfactorily reproduced.

Each manuscript should be submitted in duplicate to the Secretary of the Editorial Board, Professor Jay H. Worrell, Department of Chemistry, University of South Florida, Tampa, FL 33620. The manuscript should be typewritten in English. Nomenclature should be consistent and should follow the recommendations presented in *Nomenclature of Inorganic Chemistry*, 2nd ed., Butterworths & Co., London, 1970, and in *Pure and Applied Chemistry*, Volume 28, No. 1 (1971). Abbreviations should conform to those used in publications of the American Chemical Society, particularly *Inorganic Chemistry*.

Chemists willing to check syntheses should contact the editor of a future volume or make this information known to Professor Worrell.

TOXIC SUBSTANCES AND
LABORATORY HAZARDS

Chemicals and chemistry are by their very nature hazardous. Chemical reactivity implies that reagents have the ability to combine. This process can be sufficiently vigorous as to cause flame, an explosion, or, often less immediately obvious, a toxic reaction.

The obvious hazards in the syntheses reported in this volume are delineated, where appropriate, in the experimental procedure. It is impossible, however, to foresee every eventuality, such as a new biological effect of a common laboratory reagent. As a consequence, *all* chemicals used and *all* reactions described in this volume should be viewed as potentially hazardous. Care should be taken to avoid inhalation or other physical contact with all reagents and solvents used in procedures described in this volume. In addition, particular attention should be paid to avoiding sparks, open flames, or other potential sources that could set fire to combustible vapors or gases.

A list of 400 toxic substances may be found in the *Federal Register,* Vol. 40, No. 23072, May 28, 1975. An abbreviated list may be obtained from *Inorganic Syntheses,* Volume 18, p. xv, 1978. A current assessment of the hazards associated with a particular chemical is available in the most recent edition of *Threshold Limit Values for Chemical Substances and Physical Agents in the Workroom Environment* published by the American Conference of Governmental Industrial Hygienists.

The drying of impure ethers can produce a violent explosion. Further information about this hazard may be found in *Inorganic Syntheses*, Volume 12, p. 317. A hazard associated with the synthesis of tetramethyldiphosphine disulfide [*Inorg. Synth.,* **15,** 186 (1974)] is cited in *Inorganic Syntheses,* Volume 23, p. 199.

CONTENTS

Chapter Two INORGANIC POLYMER SYSTEMS

**Chapter Three   COMPOUNDS OF
PHARMACOLOGICAL INTEREST**

**Chapter Four   METAL COMPOUNDS, COMPLEXES,
AND LIGANDS**

Chapter Five   TRANSITION METAL ORGANOMETALLIC COMPOUNDS

CORRECTION TO VOLUME 24

DICARBONYL(η^5-CYCLOPENTADIENYL) (2-METHYL-1-PROPENYL-κC^1)IRON AND DICARBONYL(η^5-CYCLOPENTADIENYL)(η^2-2-METHYL-1-PROPENE)IRON(1+) TETRAFLUOROBORATE

SUBMITTED BY MYRON ROSENBLUM*

The preparation of dicarbonyl(η^5-cyclopentadienyl)(2-methyl-1-propenyl-κC^1)iron requires 3-chloro-2-methyl-1-propene (methallyl chloride), not the reagent given in Reference 1, p. 165.[1]

References

1. M. Rosenblum, W. P. Giering, and S.-B. Samuels, *Inorg. Synth.* **24**, 163 (1986).

*Department of Chemistry, Brandeis University, Waltham, MA 02254.

INORGANIC SYNTHESES

Volume 25

Chapter One

MAIN GROUP RING SYSTEMS AND RELATED COMPOUNDS

1. ORGANOCYCLOPHOSPHANES

Submitted by M. BAUDLER* and K. GLINKA*
Checked by A. H. COWLEY and M. PAKULSKI†

Organocyclophosphanes $(PR)_n$ $(n = 3-6)$ can be prepared by various methods, most of which start from the corresponding organodihalophosphane $RPCl_2$. The common methods of preparation[1] are either based on the dehydrohalogenation reaction between RPH_2 and $RPCl_2$ or on the dehalogenation of $RPCl_2$ with metals or metal hydrides. These reactions are nonspecific as to the ring size of the cyclophosphane formed, and in most cases those oligomers $(PR)_n$ are generated that exhibit the relatively highest thermodynamic stability. Small alkyl substituents as well as the phenyl group favor ring size five, whereas bulky organo groups lead to the formation of four-membered rings.[2] In addition, in the last decade it has

---

Editor's note: The nomenclature used in this synthesis follows accepted practice among many synthetic chemists. It differs, however, from the Chemical Abstracts recommendation in which a cyclotriphosphane would be described as a triphosphirane. A third usage describes them as cyclotriphosphines.

*Institut für Anorganische Chemie, Universität Köln, Greinstrasse 6, D-5000 Köln 41, Federal Republic of Germany.

†Department of Chemistry, University of Texas, Austin, TX 78712.

been shown that cyclotriphosphanes can be synthesized if substituents of suitable bulkiness are chosen.[3,4] The stability of these compounds must be attributed to kinetic effects.

A very interesting feature of the cyclophosphanes of various ring sizes is their ability to form complexes with metal carbonyls and transition metal salts.[1]

The procedure described for the synthesis of tri-*tert*-butyl-cyclotriphosphane, the first stable three-membered phosphorus ring compound, follows the originally published route.[5]

Pentamethylcyclopentaphosphane has been prepared by the thermal disproportionation of CH_3PF_2,[6] by the action of metallic lithium[7,8] or magnesium turnings[7] on CH_3PCl_2 and by the reaction of CH_3PBr_2 with magnesium powder.[9] The following procedure, using lithium hydride as the reducing agent, has been found advantageous because it is time saving and gives satisfactory yields.[10]

Important

Since the organodihalophosphanes and the organocyclophosphanes are easily oxidized, all reactions must be carried out under an atmosphere of dry, oxygen-free argon or nitrogen, and all reaction products must be manipulated under strict exclusion of air.

A.   TRI-*tert*-BUTYL-CYCLOTRIPHOSPHANE

$$3t\text{-}C_4H_9PCl_2 + 3Mg \longrightarrow (t\text{-}C_4H_9P)_3 + 3MgCl_2$$

Procedure

■   **Caution.**   *Careful handling of organodichlorophosphanes and organocyclophosphanes inside an efficiently working fume hood is absolutely necessary.*

A 1-L, two-necked, round-bottomed flask fitted with a lateral stopcock is equipped with a pressure-equalizing dropping funnel and a reflux condenser capped with an adapter. The lateral stopcock and the adapter are both connected via a selector valve with a source of argon (including a mercury bubbler) and a vacuum pump, respectively. The vessel is charged with 29 g (1.2 mol) of magnesium turnings and some crystals of iodine. The apparatus is evacuated and filled with argon three times. A 350-mL sample of dry tetrahydrofuran (THF) is then poured into the flask under countercurrent argon flow, and a solution of 63.6 g (0.4 mol) of *tert*-butyldichlorophosphane (*tert*-butylphosphonous dichloride)[11] in 200-mL dry THF is transferred to the dropping funnel following the same precaution as before. The flask is heated in a water bath until the solvent refluxes. After removing the bath and starting the magnetic stirrer, the t-$C_4H_9PCl_2$ solution is added dropwise at such a rate that the reflux is maintained. If no spontaneous reaction starts during the addition of the first 50 mL of the solution, it is necessary to initiate the reaction by adding some drops of bromine via the lateral stopcock using a syringe and a cannula. The reaction mixture becomes yellow with vigorous foaming, and magnesium chloride precipitates. After the addition is complete, stirring is continued for a further 15 min without heating.

The dropping funnel and the condenser are replaced with a stopper and an adapter under countercurrent argon flow via the lateral stopcock. At about 40°, the solvent is removed under vacuum (water aspirator connected with the adapter, trap cooled to $-78°$), to leave a solid, which is covered with 500 mL of dry n-pentane and crushed thoroughly with a glass rod, while argon is introduced via the lateral stopcock. The stopper is then replaced with a filtration apparatus (medium-porosity glass frit and 2-L flask, each fitted with a lateral stopcock), which previously has been evacuated and filled with argon three times, the glass frit being closed by a socket cap during these operations. The pentane suspension is poured into the filter by turning the whole apparatus upside down. Filtration proceeds by evacuating the 2-L flask via the lateral stopcock. Subsequently the filter cake is washed with 5×100-mL portions of dry n-pentane. The pentane solution is concentrated under reduced pressure to a volume of about 80 mL (water jet pump, trap cooled to $-78°$) and transferred through a bent tube to a 100-mL round-bottomed flask fitted with a lateral stopcock and filled with argon. After removing the remaining solvent by successive use of a water aspirator and an oil pump the residue is fractionated by vacuum distillation through a vacuum jacket vigreux column (15 cm) topped with a short-path distillation head that is fitted with a vertical condenser. Pure colorless $(t$-$C_4H_9P)_3$ distills at 76 to 80°/0.2 torr (bath temperature 140–160°) and solidifies in the ice-cooled receiver (yield 20.1 g, 57%).

At the end of the distillation the column may be blocked by subliming tetra-*tert*-butyl-cyclotetraphosphane, $(t$-$C_4H_9P)_4$. In this case the distillation must be interrupted at once by removing the oil bath.

Properties

The colorless needle-shaped crystals of $(t$-$C_4H_9P)_3$ (mp 41°, sealed tube) can be stored for months at $-18°$ under an inert gas atmosphere. They are readily soluble in benzene, toluene, THF, dioxane, carbon disulfide, and n-pentane, whereas contact with halogenated hydrocarbons or water brings about decomposition. The compound is rapidly attacked by atmospheric oxygen, mainly in solution. Therefore, $(t\text{-}C_4H_9P)_3$ may ignite, if it is finely divided or comes in contact with cell tissue. The ^{31}P {^1H} NMR spectrum shows an A_2B system with $\delta_A = -71.0$, $\delta_B = -109.6$, $J_{AB} = -201$ Hz (in C_6D_6); the appearance of a singlet at $\delta = -58.1$ indicates the presence of some $(t\text{-}C_4H_9P)_4$.

Reaction of $(t\text{-}C_4H_9P)_3$ with PCl_5 (molar ratio 1:2) yields the diphosphane $Cl(t\text{-}C_4H_9)P\!-\!P(t\text{-}C_4H_9)Cl$,[12] which is an excellent synthon for three-membered phosphorus heterocycles of the type $(t\text{-}C_4H_9P)_2ER_n$ (E = hetero atom, $n = 0\text{--}2$)[4] as well as for polyphosphorus compounds.[13]

B. PENTAMETHYLCYCLOPENTAPHOSPHANE

$$5CH_3PCl_2 + 10LiH \longrightarrow (CH_3P)_5 + 10LiCl + 5H_2$$

Procedure

■ **Caution.** *Pentamethylcyclopentaphosphane is extremely sensitive toward oxygen. Use the precautions listed in Section A and in addition, take into account that more than 12 L of hydrogen is evolved.*

The compound $(CH_3P)_5$ is prepared in a reaction apparatus and by a technique analogous to that described in Section A, but using a 500-mL reaction flask, a filtration apparatus with a 1-L flask, and a normal distillation link.

Over a period of 3 hr a solution of 65 g (0.56 mol) of methyldichlorophosphane (methylphosphonous dichloride)* in 100 mL of dry THF is added dropwise to a stirred suspension of 10 g (1.26 mol) of lithium hydride in 200 mL of THF. The reaction is started without external heating and the CH_3PCl_2 solution is added at such a rate, that the solvent refluxes smoothly. After additional heating at reflux for at least 3 hr, the solvent is removed *in vacuo* and the residue is crushed under 300 mL of dry *n*-pentane. The suspension is then filtered, the filter cake is washed with 5×20-mL portions of pentane, and the combined filtrates are concentrated to a volume of about 50 mL. After transferring the yellow solution to a 100-mL flask, the pentane is completely removed and the crude product is purified by vacuum distillation. Colorless $(CH_3P)_5$ distills at $100°/0.5$ torr (lit.[7] bp 110–112°/1 torr); yield 20.3 g (79%).

Properties

The compound $(CH_3P)_5$ is a colorless oily liquid with a characteristic strong odor. It is extremely sensitive to oxygen; contact with traces of air causes

*Hoechst AG, Postfach 800320, D-6230 Frankfurt am Main 80, Federal Republic of Germany, or prepared according to ref. 14.

an immediate turbidity. The compound is readily soluble in aliphatic and aromatic hydrocarbons as well as in cyclic and open-chain ethers, whereas halogenated hydrocarbons initiate decomposition. Infrared and Raman spectra of $(CH_3P)_5$ have been reported.[10] In the ^{31}P $\{^1H\}$ NMR spectrum a complex AA'BB'C pattern centered at $\delta = +19.7$ is observed.[15]

References

1. L. Maier, *Organic Phosphorus Compounds*, Vol. 1, G. M. Kosolapoff and L. Maier (eds.), Wiley-Interscience, New York, 1972, p. 339.
2. L. R. Smith and J. L. Mills, *J. Am. Chem. Soc.*, **98**, 3852 (1976).
3. M. Baudler, *Pure Appl. Chem.*, **52**, 755 (1980).
4. M. Baudler, *Angew. Chem.*, **94**, 520 (1982); *Angew. Chem. Int. Ed.* (*Engl.*), **21**, 492 (1982).
5. M. Baudler, J. Hahn, H. Dietsch, and G. Fürstenberg, *Z. Naturforsch.*, **B 31**, 1305 (1976); M. Baudler and Ch. Gruner, *Z. Naturforsch.*, **B 31**, 1311 (1976).
6. V. N. Kulakova, Yu.M. Zino'ev, and L. Z. Soborovskiĭ, *Zh. Obshch. Khim.*, **29**, 3957 (1959); *Chem. Abs.* **54**, 20846e (1960).
7. Wm. A. Henderson, Jr., M. Epstein, and F. S. Seichter, *J. Am. Chem. Soc.*, **85**, 2462 (1963).
8. K. Issleib, Ch. Rockstroh, I. Duchek, and E. Fluck, *Z. Anorg. Allg. Chem.*, **360**, 77 (1968).
9. W. Kuchen and W. Grünewald, *Chem. Ber.*, **98**, 480 (1965).
10. M. Baudler and K. Hammerström, *Z. Naturforsch.*, **B 20**, 810 (1965).
11. M. Fild, O. Stelzer, and R. Schmutzler, *Inorg. Synth.*, **14**, 4 (1973).
12. M. Baudler, J. Hellmann, and J. Hahn, *Z. Anorg. Allg. Chem.*, **489**, 11 (1982).
13. M. Baudler and W. Göldner, *Chem. Ber.*, **118**, 3268 (1985).
14. L. Maier, *Inorg. Synth.*, **7**, 82 (1963), note 2.
15. J. P. Albrand, D. Gagnaire, and J. B. Robert, *J. Am. Chem. Soc.*, **95**, 6498 (1973); **96**, 1643 (1974).

2. DIETHYLAMMONIUM cyclo-OCTATHIOTETRAPHOSPHATE(III)

$$P_4 + 2(Et_2NH_2)_2S_4 \longrightarrow (Et_2NH_2)_4(P_4S_8)$$

Submitted by HANS H. FALIUS* and ANDREAS B. SCHLIEPHAKE*
Checked by A. H. COWLEY and M. PAKULSKI†

Compounds containing phosphorus–phosphorus bonds are synthesized either by coupling of phosphorus atoms or by partial chemical cleavage of

*Institut für Anorganische und Analytische Chemie, Technische Universität, Hagenring 30, 3300 Braunschweig, Federal Republic of Germany.
†Department of Chemistry, University of Texas, Austin, TX 78712.

the bonds in elemental phosphorus.[1] Normally the latter method does not lead to uniform reaction products, but compounds that have not yet been prepared by other methods are isolable from the mixtures.[2,3] White phosphorus readily reacts with polysulfidic sulfur and to a large extent only two bonds are broken in the P_4 tetrahedra. This leads to the first known square homocycle consisting of four tetracoordinated phosphorus atoms. The reaction is rapid and the cyclic tetraphosphate can be isolated in 2 hr giving 50% yield.

■ **Caution.** *Because of the very poisonous nature of the H_2S, the reaction must be performed in an efficient hood. Notice should be taken of the usual precautions for handling white phosphorus: It is very air sensitive and must be stored and cut under water. Before it is weighed and used in the reaction, it is washed with methanol. Protective gloves, glasses, and clothing should be worn. Low valent phosphorus compounds are toxic, and contact with eyes or mucous membranes should be strictly avoided.*

Procedure

Solvents and H_2S are dried (*N-N*-dimethylformamide with 4-A molecular sieves, diethylamine with sodium wire, and hydrogen sulfide with P_4O_{10}). To a three-necked, round-bottomed, 250-mL flask equipped with a magnetic stirrer, thermometer, reflux condenser, and a gas inlet (vented through a safety bubbler) is added 60 mL (0.779 mol) of *N,N*-dimethylformamide (DMF), 25.1 mL (0.242 mol) of diethylamine, and 10.35 g (0.323 mol) of sulfur. The mixture is heated in a water bath to 45° and flushed with hydrogen sulfide with vigorous stirring until the solution is saturated (about 20–30 min). The hydrogen sulfide source is then removed and replaced with nitrogen. After addition of 5.00 g (0.161 mol) of white phosphorus, the dark brown solution is stirred slowly. Once the phosphorus has melted the reaction begins and the reaction rate depends on the degree of distribution of the molten phosphorus. If the exothermic reaction proceeds too quickly the temperature will rise above 50° resulting in a greater amount of by-products. When the evolution of a small amount of gas has ended, the solution can be stirred faster. After the product has precipitated and the solution has turned dark green, the mixture is stirred vigorously for 20 min in order to be certain that no elemental phosphorus remains unchanged. The total reaction time after addition of phosphorus is about 1 hr. After cooling to room temperature, the precipitate is removed by filtering with a Büchner funnel and washed with DMF until it is white. Then it is suspended in acetonitrile, filtered by suction again, washed with acetonitrile and ether, and dried under vacuum (20°, 0.01 torr, 20 min) to give 12.3–14.2 g (45–52% yield) of $(Et_2NH_2)_4(P_4S_8)$.

Anal. Calcd. for $C_{16}H_{48}N_4P_4S_8$: C, 28.4; H, 7.2; N, 8.3; P, 18.3; S, 37.9. Found: C, 28.3; H, 7.3; N, 8.2; P, 18.5; S, 37.5. The material so prepared is generally pure enough for most purposes.

Properties

The *cyclo*-tetraphosphate is a slightly yellow crystalline solid, mp 144°, which is readily soluble in water and formamide. The compound is stable in air and the anion is hydrolyzed very slowly.[4] Its ^{31}P NMR spectrum (formamide) shows a singlet at δ = 121.6 ppm (85% H_3PO_4, external) and the IR spectrum has been reported.[5] The diethylammonium salt may easily be converted to the ammonium salt by pouring its aqueous solution into a concentrated solution of ammonium nitrate. The compound $(NH_4)_4(P_4S_8) \cdot 2H_2O$ forms less soluble shiny flaky platelets. The $[P_4S_8]^{4-}$ anion contains a square planar P_4 ring[4] and gives precipitates with Ag, Cu^{II}, Cd, Hg^{II}, Sn^{II}, Pb^{II}, and Bi^{III} cations.

References

1. A. H. Cowley, *Compounds Containing Phosphorus–Phosphorus Bonds*, Benchmark Papers, Dowden, Hutchinson, and Ross Inc., Stroudsburg, PA, 1973.
2. H. H. Falius, *Z. Anorg. Allgem. Chem.*, **394,** 217 (1972); **396,** 245 (1973).
3. W. Krause and H. H. Falius, *Z. Anorg. Allgem. Chem.*, **496,** 80 (1983).
4. H. H. Falius, W. Krause, and W. S. Sheldrick, *Angew. Chem.*, **93,** 121 (1981); *Angew. Chem. Int. Ed. (Engl.)*, **20,** 103 (1981).
5. H. Bürger, G. Pawelke, and H. H. Falius, *Spectrochim. Acta*, **37 A,** 753 (1981).

3. TERVALENT PHOSPHORUS–NITROGEN RING COMPOUNDS

Submitted by O. J. SCHERER* and R. ANSELMANN*
Checked by R. T. PAINE† and S. KARTHIKEYAN†

The first cyclodiphosphazane (diazadiphosphetidine) was reported by Michaelis and Schröter[1] in 1894. The synthesis of 1,3-di-*tert*-butyl-2,4-di-

*Fachbereich Chemie, Universität Kaiserslautern, Erwin-Schrödinger-Strasse, D-6750 Kaiserslautern, Federal Republic of Germany.
†Department of Chemistry, The University of New Mexico, Albuquerque, NM 87131.

chlorocyclodiphosphazane, $(ClPNCMe_3)_2$, in 1969[2] and later,[3,4] the X-ray sturcture characterization of the cis isomer,[5] as well as the discovery of monomeric aminoiminophosphanes[6] of the type $RR'N—P{=}NR''$ gave important impulses to the chemistry of these four-membered tervalent phosphorus-nitrogen ring compounds.[7] Chlorine atom substitution reactions, oxidation of one or both phosphorus atoms,[7] separation of cis and trans isomers, and especially coordination chemistry[4,7,8] have been studied intensively.

A.   1,3-DI-*tert*-BUTYL-2,4-DICHLOROCYCLODIPHOSPHAZANE

Method 1

$$2(Me_3Si)(Me_3C)NLi + 2PCl_3 \longrightarrow cis\text{-} \quad \begin{array}{c} \text{(ring structure)} \end{array} \quad + 2LiCl + 2Me_3SiCl$$

Procedure a: $(Me_3Si)(Me_3C)NH$ $(Me_3Si)(Me_2CH)NH$

The reaction vessel is a 2-L, three-necked flask equipped with a mechanical stirrer, a water-cooled reflux condenser, and a 1-L pressure-equalizing dropping funnel. Under inert atmosphere conditions, distilled chlorotrimethylsilane (120 mL, 102.7 g, 0.95 mol) in diethyl ether (500 mL) is added dropwise at room temperature from the dropping funnel to an efficiently stirred solution of *tert*-butylamine (100 mL, 69.6 g, 0.95 mol, freshly distilled from KOH) and triethylamine (135 mL, 98.2 g, 0.97 mol, freshly distilled from KOH) in diethyl ether (500 mL). The reaction mixture is stirred for 2 hr. Solid amine hydrochloride is filtered off (D3 frit) and washed twice with 100 mL of diethyl ether. Fractional distillation yields 100–120 g (70–80%), bp 119–122°/760 torr.

The reaction between 0.95 mol of chlorotrimethylsilane and isopropylamine (162 mL, 118.2 g, 2.0 mol) affords 87–112 g (70–80%), bp 100–104°/760 torr.

Properties

The properties of trimethylsilyl-*tert*-butylamine[9] [^{1}H NMR (C_6H_6, TMS int.): $\delta_{CH_3Si} = 0.25$ ppm(s), $\delta_{CH_3C} = 1.2$ ppm(s)], and trimethylsilyl-isopropylamine[10] [^{1}H NMR (C_6H_6, TMS int.): $\delta_{CH_3Si} = 0.10$ ppm(s), $\delta_{CH} = 2.95$ ppm(d, sp), $\delta_{CH_3} = 1.0$ ppm(d), $^3J_{HH} = 6.9$ Hz] have been reported.

Procedure b: [*ClPNCMe₃*]₂

A solution of 146 mmol (Me₃Si)(Me₃C)NLi [prepared from 21.2 g (146 mmol) (Me₃Si)(Me₃C)NH in 100 mL of diethyl ether and 146 mmol of *n*-butyllithium in hexane (commercially available from Metallgesellschaft as ~1.6 *M* LiBu in hexane) with magnetic stirring and 1-hr reflux] is added dropwise at room temperature with magnetic stirring to 21.0 g (153 mmol) of PCl₃ in 200 mL of diethyl ether. The reaction mixture is then stirred for 30 min and reduced in volume to approximately 200 mL (half). Lithium chloride is filtered off (D4 frit) and washed with pentane. The solvent from the combined filtrates is removed by room temperature evaporation under reduced pressure. The residue is heated to 130° (oil bath) while the pressure is slowly reduced to 0.01 torr [bp 70–75°/0.01 torr, mp 42–44°, yields varying from 13.4–15.5 g (65–75%].

Method 2

$$2PCl_3 + 6Me_3CNH_2 \longrightarrow cis\text{-}[ClPNCMe_3]_2 + 4Me_3CNH_2 \cdot HCl$$

tert-Butylamine (62.9 g, 0.860 mol) is added dropwise to an efficiently stirred solution of PCl₃ (39.4 g, 0.287 mol) in diethyl ether (600 mL) at −78°. On completion of the addition, the mixture is allowed to warm to room temperature and is stirred overnight. Solid amine hydrochloride is removed by filtration and the solid is carefully washed with diethyl ether (200 mL), and the washings are added to the filtrate. The solvent is removed by room temperature evaporation under reduced pressure. Traces of diethyl ether and a considerable quantity of Cl₂PNHCMe₃ are finally removed at 35–75°/0.1 torr. Distillation at 95°/0.1 torr followed either by sublimination (50–70°/0.05–0.001 torr) or recrystallization from light petroleum (bp 40–60°) gives *cis*-[ClPNCMe₃]₂ in yields varying from 35 to 63%, mp 42–43°. The product is completely free from 1,3-di-*tert*-butyl-2-chloro-4-*tert*-butylamino-cyclodiphosphazane.

Properties

cis-[ClPNCMe₃]₂ is an air and moisture sensitive, colorless solid, mp 42–44° ¹H NMR (C₆H₆, TMS int.): δ = 1.18 ppm(t), $^4J_{HH}$ = 1.1 Hz. ³¹P {¹H} NMR (C₆H₆, 85% H₃PO₄ ext.): δ = 207.8 ppm(s).

B. 2,4,6,8,9,10-HEXA(ISOPROPYL)-2,4,6,8,9,10-HEXAAZA-1λ³,3λ³,5λ³,7λ³-TETRAPHOSPHATRICYCLO-[5.1.1.1.³,⁵]DECANE

Cyclodiphosphazanes (diazadiphosphetidines) are very useful starting materials for cages and cagelike (polycyclic) P—N compounds. The eight-membered cage compound $P_4(NCMe_3)_4$ is formed by the reduction of *cis*-$[ClPNCMe_3]_2$ with Mg.[11] The compound $P_4(NCHMe_2)_6$ is a nonadamantane-type molecule that contains two superimposed diazadiphosphetidine rings coupled through the phosphorus atoms by bridging $NCHMe_2$ groups and has been synthesized according to the reaction sequence in eqs. (1) to (4).[12]

$$(Me_3Si)(Me_2\overset{\frown}{CH})NLi + PCl_3 \longrightarrow Cl_2PN(CHMe_2)(SiMe_3) + LiCl \quad (1)$$

$$2Cl_2PN(CHMe_2)(SiMe_3) \overset{\Delta}{\longrightarrow} [ClPNCHMe_2]_2 + 2Me_3SiCl \quad (2)$$

$$[ClPNCHMe_2]_2 + (MeSi)(Me_2CH)NLi \longrightarrow$$
$$(Me_3Si)(Me_2CH)N\overline{PNCHMe_2PClN}CHMe_2 + LiCl \quad (3)$$

$$2(Me_3Si)(Me_2CH)N\overline{PNCHMe_2PClN}CHMe_2 \overset{\Delta}{\longrightarrow}$$
$$P_4(NCHMe_2)_6 + 2Me_3SiCl \quad (4)$$

Procedure a: 2,4-Dichloro-1,3-diisopropylcyclodiphosphazane,
$[ClPNCHMe_2]_2$

To a solution of 70.04 g (0.51 mol)PCl_3 in 250 mL of diethyl ether a solution of 0.50 mol of $(Me_3Si)(Me_2CH)NLi$ [prepared from 65.65 g (92.5 mL, 0.5 mol) $(Me_3Si)(Me_2CH)NH$ in 300 mL of diethyl ether and 0.5 mol of *n*-butyllithium in hexane] is added dropwise (2 hr) at $-78°$ with magnetic stirring. After removal of the dry ice–acetone bath the reaction mixture is stirred overnight and reduced in volume under vacuum to approximately 250 mL. Lithium chloride is filtered off (D4 frit) and washed with pentane (100 mL). The filtrate is again reduced (vacuum) in volume to approximately 200 mL and heated (water-cooled reflux condenser) for 2 hr at 50 to 60° (oil bath temperature). Chlorotrimethylsilane and the solvent are removed in vacuum. The residue is fractionally distilled [bp 45–50°/0.001 torr, yield 45 g (73%)].

Properties

2,4-Dichloro-1,3-diisopropylcyclodiphosphazane is an air- and moisture-sensitive, colorless liquid.[3] ^{1}H NMR (C_6H_6, TMS int.): $\delta_{CH} = 3.39$ ppm(t,sp), $^3J_{PH} = 0.9$ Hz, $^3J_{HH} = 6.5$ Hz; $\delta_{CH_3} = 1.13$ ppm(d,t), $^4J_{PH} = 0.9$ Hz, $^3J_{HH} = 6.4$ Hz. ^{31}P $\{^1H\}$ NMR (C_6H_6, 85% H_3PO_4 ext.): $\delta = 220.6$ ppm(s). The checkers point out that this compound is not as stable as the *tert*-butyl analog. It decomposed within 24 hr even in rigorously dried glassware at room temperature. It is stable, however, at $-78°$ for at least a week.

Procedure b: 2-Chloro-1,3-diisopropyl-4-isopropyltimethylsilylamino-
1,3,2λ^3,4λ^3-cyclodiphosphazane, $(Me_3Si)(Me_2CH)$
$N\overline{PNCHMe_2PClN}CHMe_2$

To a solution of 25.27 g (102.3 mmol) of [ClPNCHMe$_2$]$_2$ in 50 mL of diethyl ether a solution of 102.3 mmol of (Me$_3$Si)(Me$_2$CH)NLi [prepared from 13.42 g (18.9 mL, 102.3 mmol) of (Me$_3$Si)(Me$_2$CH)NH in 50 mL of diethyl ether and 102.3 mmol of *n*-butyllithium in hexane] is added dropwise (30 min) with magnetic stirring at $-16°$ (ice–sodium chloride mixture). [*Note:* The stoichiometry must be exact: even a small excess of LiN(*i*-Pr)TMS induces decomposition.] After removal of the cold bath the reaction mixture is stirred for 45 min at room temperature and reduced in volume to approximately 50 mL. Lithium chloride is filtered off (D4 frit) and washed twice with 10 mL of pentane. The solvent is removed under reduced pressure (0.01 torr) to yield 33.1 g (95%) of an orange-brown oil (crude).

Properties

Crude (Me$_3$Si)(Me$_2$CH)N$\overline{\text{PNCHMe}_2\text{PClN}}$CHMe$_2$ is an exceedingly air- and moisture-sensitive, orange-brown oil (NMR spectroscopically measured purity >90%). ^{31}P {^{1}H} NMR (C$_6$H$_6$, 85% H$_3$PO$_4$ ext., 293 K): $\delta_{P(N)} = $ 185.4(cis), 208.5(trans), $\delta_{P(Cl)} = $ 130.2(cis), 175.8 ppm(trans); cis isomer: trans isomer = 80: 20%. Beause of the large half-width at 293 K all signals are broad singlets (C$_7$D$_8$/203 K: all signals are doublets $^2J_{PP} = $ 45.8(cis) and 42.7 Hz(trans). It should be used immediately, since it decomposed within 12 hr, even at $-78°$.

Procedure c: 2,4,6,8,9,10-Hexa(isopropyl)-2,4,6,8,9,10-hexaaza-1λ^3,3λ^3,5λ^3,7λ^3-tetraphospha-tricyclo[5.1.1.1^{3,5}]decane, P$_4$(NCHMe$_2$)$_6$

A 9.65 g (28 mmol) sample of (Me$_3$Si)(Me$_2$CH)N$\overline{\text{PNCHMe}_2\text{PClN}}$CHMe$_2$ (crude) in a 100-mL Schlenk tube is dissolved in 50 mL of acetonitrile and refluxed for 2 hr. After cooling to room temperature slightly yellow-orange crystals precipitate and were separated. With the filtrate this procedure is repeated three times. After the fourth refluxing period the solution is cooled to $-35°$. The collected crystals yield 2.7 g (41%) of crude product with approximately 10% "impurity" of the adamantane-type isomer. Recrystallization from 60 mL of hot acetonitrile and cooling slowly to room temperature affords 2.2 g (34%) of P$_4$(NCHMe$_2$)$_6$. [*Note:* The checkers experienced some difficulties with this synthesis and obtained much lower yields (~7%).] A key factor appears to be the length of time that passes between synthesis and reaction of the trimethylsilyl precursor.

Properties

P$_4$(NCHMe$_2$)$_6$ forms air- and moisture sensitive colorless crystals [12]. The crystal structure confirms the rectangular arrangement of the four phosphorus atoms [12]. ^{1}H NMR (CD_2Cl_2, TMS int.): δ CH_3(bridge) = 1.19 ppm(d), $^3J_{HH}$ = 6.8 Hz; δ CH_3(four-membered ring) = 1.20 ppm(d), $^3J_{HH}$ = 6.4 Hz, δ CH = 3.82 – 4.22 ppm(m). ^{31}P{^{1}H}NMR (C_6H_6, 85% H_3PO_4 ext.) δ = 147.0 ppm(s) [12]. Thermolysis (156 – 158°C, several days) yields the adamantane-type isomer. ^{1}H NMR (C_6H_6, TMS int.): δ CH_3 = 1.37 ppm(d), $^3J_{HH}$ = 6.8 Hz. ^{31}P{^{1}H} NMR (C_6H_6, 85% H_3PO_4 ext.): δ = 84.0 ppm(s) [12]. The oxidation for example with sulfur yields $P_4(NCHMe_2)_6S_x$ (x = 1,2; only the isomer where two diagonal arranged phosphorus atoms are oxidized) [13]. The ligating properties (P-coordination) have been realized with the synthesis of the complexes $[P_4(NCHMe_2)_6\text{-}Cr(CO)_5]$ and $[P_4(NCHMe_2)_6]_2$ Au^+Cl^- [14].

References

1. A. Michaelis and G. Schröter, *Ber. Dtsch. Chem. Ges.*, **27**, 490 (1894).
2. O. J. Scherer and P. Klusmann, *Angew. Chem.*, **81**, 743 (1969); *Angew. Chem. Int. Ed. (Engl.)*, **8**, 752 (1969).
3. R. Jefferson, J. F. Nixon, T. M. Painter, R. Keat, and L. Stobbs, *J. Chem. Soc. Dalton Trans.*, **1973**, 1414.
4. J. C. T. R. Burckett, St. Laurent, P. B. Hitchcock, and J. F. Nixon, *J. Organomet. Chem.*, **249**, 243 (1983).
5. K. W. Muir and J. F. Nixon, *Chem. Commun.*, **1971**, 1405; K. W. Muir, *J. Chem. Soc. Dalton Trans.*, **1975**, 259.
6. Reviews: (a) E. Niecke and O. J. Scherer, *Nachr. Chem. Tech.*, **23**, 395 (1975). (b) E. Fluck, *Top. Phosphorus Chem.*, **10**, 194 (1980).
7. Reviews: (a) A. F. Grapov, N. N. Mel'nikov, and L. V. Razvodovskaya, *Russ. Chem. Rev.*, **39**, 20 (1970). (b) O. J. Scherer, *Nachr. Chem. Tech. Lab.*, **28**, 6 (1980). (c) R. Keat, *Top. Curr. Chem.*, **102**, 89 (1982).
8. O. J. Scherer, R. Anselmann, and W. S. Sheldrick, *J. Organomet. Chem.*, **263**, C26 (1984).
9. R. M. Pike, *J. Org. Chem.*, **26**, 232 (1961), E. C. Ashby, J. J. Lin, and A. B. Goel, *J. Org. Chem.*, **43**, 1564 (1978).
10. A. W. Jarvie and D. Lewis, *J. Chem. Soc.*, **1963**, 1073.
11. D. Du Bois, E. N. Duesler, and R. T. Paine, *J. Chem. Soc., Chem. Commun.*, **1984**, 488.
12. O. J. Scherer, K. Andres, C. Krüger, Y.-H. Tsai, and G. Wolmershäuser, *Angew. Chem.*, **92**, 563 (1980); *Angew. Chem., Int. Ed. (Engl.)*, **19**, 571 (1980).
13. K. Andres, Ph.D. Thesis, University of Kaiserslautern, 1983.
14. O. J. Scherer and R. Anselmann, unpublished results.

4. 2,4,6-TRICHLORO-1,3,5-TRIETHYLCYCLOPHOSPH(III)AZANES (1,3,5,2,4,6-TRIAZATRIPHOSPHORINANES)

$$3PCl_3 + 3NEtH_3\,Cl \longrightarrow \quad \underset{\substack{\\ Cl}}{\overset{\substack{Et \\ N}}{\underset{Et}{\overset{Cl}{\big|}}\!\diagup\!\!\diagdown\!\overset{Cl}{\diagdown}\!\diagup\!\overset{N}{\underset{Et}{}}}} \quad + 9HCl$$

Submitted by JAMES GEMMILL* and RODNEY KEAT*
Checked by RAHIM HANI† and ROBERT H. NEILSON†

Introduction

The cyclotriphosph(III)azanes (triazatriphosphorinanes) $(XPNR)_3$ (R = alk, X = Cl) have, as yet, received little attention,[1] mainly because a convenient route to synthesis is lacking. The synthesis[2] of $(ClPNMe)_3$ from PCl_3 and $(Me_3Si)_2NMe$ requires a source of the latter compound, and this method does not give good yields of the P_3N_3-ring compound. Recent work[3] shows that the analogous ethyl compound, $(ClPNEt)_3$ (but not the methyl compound) can be prepared in good yield using readily available materials, and the method is practical on a large scale. The chemistry of this compound is beginning to be explored, and it offers interesting possibilities for the study of the properties of $P_3^{(III)}N_3$ ring compounds, including those of a potentially tridentate phosphorus compound. The product is a mixture of cis and trans isomers, with the latter predominating at ambient temperatures. Care must be taken to avoid contact of the hot reaction mixture and the hot product with air. All ground glass joints should be thoroughly coated with a silicon or, preferably, Kel-F grease.

■ **Caution.** *If the initial reflux described here stops, and the reaction mixture cools, the $NEtH_3Cl$ will form a solid surface crust. This crust must be broken before any attempt is made to restart the reaction, otherwise there will be a violent "bump" and subsequent foaming as the solvent breaks the crust. Glass joints can easily be blown apart by this action, and a serious fire–toxicity hazard may ensue. The synthesis should be carried out in a fume hood because of the HCl evolution. 1,1,2,2-Tetrachloroethane is a toxic solvent.*

*Department of Chemistry, University of Glasgow, Glasgow G12 8QQ, Scotland, United Kingdom.
†Department of Chemistry, Texas Christian University, Fort Worth, TX 76129.

Procedure

Phosphorus trichloride (220 g, 1.6 mol), ethanamine hydrochloride (98 g, 1.2 mol), and 1,1,2,2-tetrachloroethane (1 L, freshly distilled from P_4O_{10}) are mixed in a 3-L round-bottomed flask fitted with a 50-cm dimpled air condenser topped with a 30-cm water condenser. The top of the water condenser is fitted with an inlet–outlet adaptor (Dreschel type). The inlet is connected to a supply of dry nitrogen and the outlet is connected to a paraffin–oil bubbler. Antibumping stones are added to the flask. The apparatus is flushed with nitrogen for 10 min. Then the gas flow is stopped, and the mixture is boiled under reflux, using a 3-L heating mantle for 3 to 4 days. The 1,1,2,2-tetrachloroethane is allowed to recondense no higher than halfway up the air condenser. Hydrogen chloride gas is evolved, and a little phosphorus(III) trichloride is often lost despite the condenser arrangement. This loss can be minimized by careful control of the heating over the first few hours. When the evolution of gas has stopped, the mixture, now a red-brown solution, is allowed to cool under a steady stream of nitrogen gas. The air and water condensers are replaced with an air condenser (50 cm) fitted with distillation attachments and connected to a 2-L round-bottomed receiving flask. The apparatus is flushed with nitrogen for 10 min, and the 1,1,2,2-tetrachloroethane is distilled off, under nitrogen, until the solution has a volume of approximately 250 mL. The apparatus is once again cooled under a flow of nitrogen and the thick red-brown solution is transferred to a 500-mL round-bottomed flask that is connected to a vacuum line by thick walled rubber tubing via a trap, cooled by solid CO_2, to collect the remaining solvent. The remaining 1,1,2,2-tetrachloroethane is then removed by heating to 80° under reduced pressure while the solution is stirred by a magnetic stirring bar. When the last traces of solvent have been removed, the apparatus is allowed to cool to room temperature. The thick red oil is now transferred to the smallest possible round-bottom flask, normally 250 mL, and is distilled at 0.05 torr at 135 to 140° (checkers: 117–122° at 0.1 torr) using a short air condenser. The product is a clear mobile oil that slowly crystallizes in the receiving vessel (mp 25–30°, yield 90 g, 70%).

Anal. Calcd. for $C_6H_{15}Cl_3N_3P_3$: C,21.9; H,4.6; N,12.8. Found: C,21.8; H,5.1; N,12.9.

At temperatures above 145° a variable amount of a second product is obtained from the distillation. This is $Et_5N_5P_4Cl_2$.[3] It appears as a viscous heavy yellow oil that quickly solidifies.

Properties

The product is a colorless, crystalline, low melting solid. It is air and water sensitive, particularly at elevated temperatures, but it can be handled in air at room temperature for brief periods. It may ignite spontaneously in the presence of water. Long term storage should be under nitrogen in a stoppered flask.

Phosphorus-31 NMR spectroscopy is the best method of checking identity and purity. The product, in $CDCl_3$, shows a singlet at 104.1 ppm for the cis isomer and a 2:1 doublet:triplet at 134.5 and 129.0 ppm, respectively, for the trans isomer, J_{PNP} unresolved at room temperature.[3]

The most common impurities are $Et_5N_5P_4Cl_2$ and the oxides of (Et-NPCl)$_3$. These are easily identified[3] by ^{31}P NMR spectroscopy and the oxides give signals well upfield of those noted here. The oxides may be removed by recrystallization from light petroleum (bp 40–60°).

References

1. R. Keat, *Top. Curr. Chem.*, **102,** 89 (1982).
2. W. Zeiss and K. Barlos, *Z. Naturforsch.*, **34B,** 423 (1979).
3. D. A. Harvey, R. Keat, and D. S. Rycroft, *J. Chem. Soc. Dalton Trans.*, **1983,** 425.

5. AMINOCYCLOTETRAPHOSPHAZENES AND TRANSANNULAR BRIDGED BICYCLIC PHOSPHAZENES

Submitted by A. C. SAU*, K. S. DHATHATHREYAN*, P. Y. NARAYANASWAMY,* and S. S. KRISHNAMURTHY*
Checked by J. C. VAN DE GRAMPEL†, A. A. VAN DER HUIZEN†, and A. P. JEKEL†

Aminolysis reactions of $1,3,5,7,2\lambda^5,4\lambda^5,6\lambda^5,8\lambda^5$-tetraazatetraphosphocine (octachlorocyclotetraphosphazene)‡($N_4P_4Cl_8$) with primary amines, as well as the reactions of bis(primary amino) hexachloro derivatives, $N_4P_4Cl_6(NHR)_2$, with secondary amines proceed in three distinct pathways: (a) a stepwise replacement of chlorine atoms to yield partially and fully substituted cyclotetraphosphazenes(tetraazatetraphosphocines), (b) inter-

Editor's Note: Nearly all investigators working in this area use the "phosphazene" based nomenclature (e.g., cyclotetraphosphazene) rather than the Chemical Abstracts notation. For completeness, both names are given. Note, however, that numbering of the ring atoms begins at nitrogen in the Chemical Abstracts terminology but at phosphorus in the phosphazene notation.

*Department of Inorganic and Physical Chemistry, Indian Institute of Science, Bangalore-560 012, India.

†Department of Inorganic Chemistry, University of Groningen, Nijenborgh 1G, 9747 AG, Groningen, The Netherlands.

molecular condensation reactions that lead to the formation of resins, and (c) an intramolecular substitution reaction to afford novel transannular bridged bicyclic phosphazenes. The relative yields of (amino)cyclotetraphosphazenes and bicyclic phosphazenes depend on the solvent, the primary amino substituent present on the phosphazene substrate, and the attacking nucleophile. A proton abstraction mechanism that involves the intermediacy of a three-coordinated P^V species has been invoked to explain the formation of the transannular P—N—P bridge. (Amino)cyclotetraphosphazenes and bicyclic phosphazenes can be readily distinguished by their thin layer chromatography (TLC) R_f values, IR data, and ^{31}P NMR spectroscopy. The $^{31}P \{^1H\}$ NMR spectra of these derivatives constitute excellent examples of different types of a four spin system.[1-7] The molecular skeleton of the bicyclic phosphazenes look somewhat like adamantane.[8] (Amino)cyclophosphazenes are of considerable interest in view of their ability to function as versatile ligands in transition metal complexes.[9]

The syntheses of a few (amino)cyclotetraphosphazenes and transannular bridged bicyclic phosphazenes derived from them are described next.

■ **Caution.** *Owing to the high volatility of the alkyl amines, all the reactions must be carried out in an efficient fume hood.*

A. 2,4,4,6,8,8-HEXACHLORO-*Trans*-2,6-BIS(ETHYLAMINO)-1,3,5,7,2λ⁵,4λ⁵,6λ⁵,8λ⁵-TETRAAZATETRAPHOSPHOCINE [1,3,3,5,7,7-HEXACHLORO-*Trans*-1,5-BIS(ETHYLAMINO)CYCLOTETRAPHOSPHAZENE] (II)

Procedure

A three-necked, 500-mL, round-bottomed flask equipped with a glass-jacketed addition funnel and a potassium hydroxide drying tube is placed in a cooling bath containing ice–water slurry. A magnetic stirring bar is introduced, and the flask is charged with 9.28 g (0.02 mol) of octachlorocyclotetraphosphazene (octachloro-1,3,5,7,2λ^5,4λ^5,6λ^5,8λ^5-tetraazatetraphosphocine)($N_4P_4Cl_8$) (I),* followed by the addition of 200 mL of anhydrous diethyl ether (distilled and preserved over sodium wire). This neck is closed with a ground glass stopper and the contents of the flask are stirred until the crystals dissolve. Anhydrous sodium sulfate (50 g) is added to the solution. Ethylamine solution (50% w/v) in water (8.0 mL, equivalent to 0.08 mol of ethylamine)† is placed in the dropping funnel, cooled by placing crushed ice in the jacket of the addition funnel, and is added to the ethereal solution of $N_4P_4Cl_8$ with vigorous stirring‡ over a period of 45 min. The reaction mixture is stirred at 0° for 1.5 hr and then allowed to attain room temperature (25°). Ethylamine hydrochloride and sodium sulfate are removed by filtration and washed twice with 25 mL of diethyl ether. The filtrate and the washings are combined. Removal of the solvent from the combined solution, using a rotary flash evaporator, yields an oil that is extracted with 100 mL of boiling petroleum (bp 60–80°). The solution is filtered while hot, concentrated to 50 mL, and cooled to 25°. Two crops of crystals (mp 109–112°) deposited at intervals of 1 hr are removed by filtration. The mother liquor is concentrated and cooled to obtain two more crops of the same crystalline material. Recrystallization of the crude product from petroleum gives pure *trans*-2,6 bis(ethylamino) derivative (II) (5.3 g, 55%).

Anal. Calcd. for $C_4H_{12}Cl_6N_6P_4$: C 10.0; H 2.5; N 17.5; Cl 44.3. Found: C 10.1; H 2.6; N 17.3; Cl 44.0.

Properties

The colorless crystals of compound II melt at 116°. The compound can be stored in a desiccator for a few days, without any appreciable decompo-

---

*A sample of $N_4P_4Cl_8$ can be obtained from Ethyl Corporation, Baton Rouge, LA 70989, or can be prepared.[10] It is recrystallized from petroleum (bp 60–80°) prior to use.

†Anhydrous ethylamine, dissolved in Et_2O can also be employed; in that case, the use of Na_2SO_4 is unnecessary. Aqueous solution of ethylamine is more readily available commercially.

‡The checkers prefer the use of a mechanical stirrer.

sition and can be preserved in sealed ampules for longer periods. The IR spectrum of the compound shows a strong broad band centered at 1304 cm^{-1} attributable to a ring P=N stretching vibration. The ^1H NMR spectrum shows the CH_2 and CH_3 resonances at $\delta = 3.12$ and 1.26. The ^{31}P $\{^1H\}$ NMR spectrum is of the A_2B_2 type; $\delta_{PCl_2} = -3.4$, $\delta_{PClNHEt} = -4.9$, $^2J_{AB} = 46.0$ Hz.[11]

B. 3,3,5,7,7-PENTAKIS(DIMETHYLAMINO)-9-ETHYL-1-ETHYLAMINO-2,4,6,8,9-PENTAAZA-1λ^5,3λ^5,5λ^5,7λ^5-TETRAPHOSPHABICYCLO[3.3.1]-NONA-1,3,5,7-TETRAENE (III)

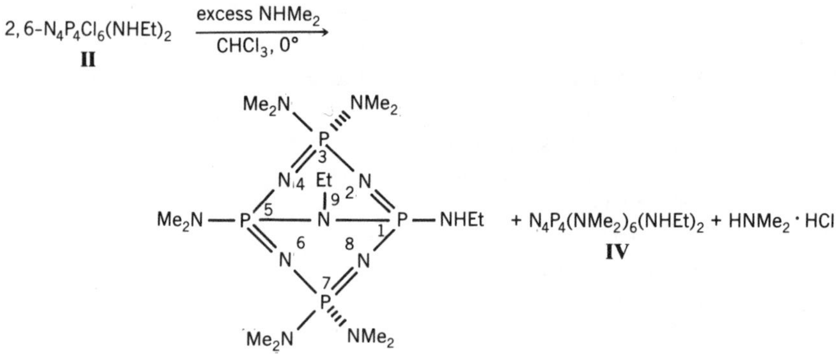

$2,6-N_4P_4Cl_6(NHEt)_2$ $\xrightarrow[\text{CHCl}_3,\,0°]{\text{excess NHMe}_2}$

II

$+ N_4P_4(NMe_2)_6(NHEt)_2 + HNMe_2 \cdot HCl$

IV

III

Procedure

Compound II, obtained by Procedure A, (4.8 g, 0.01 mol) is dissolved in 250 mL of chloroform (freshly distilled over P_4O_{10} and free from ethanol) in a 500-mL, two-necked, round-bottomed flask fitted with a condenser maintained at $-75°$ (acetone–solid CO_2), an addition funnel, and a magnetic stirring bar. Ingress of moisture is prevented by placing a potassium hydroxide guard tube at the top of the condenser. The phosphazene solution is cooled in an ice–water slurry. Anhydrous dimethylamine (22.5 g, 0.50 mol)* is added rapidly to the phosphazene solution under vigorous stirring. The reaction mixture is stirred for 1 hr at $0°$ and for 3 hr at $\sim25°$. Solvent is removed using a rotary evaporator and the residual viscous oil is treated with 50 mL of hot petroleum (bp 60–80°). The insoluble dimethylamine hydrochloride is removed by filtration. The filtrate is concentrated

---

*The ampule containing Me_2NH has to be cooled to $-25°$ before opening; the dropping funnel has to be cooled at the same temperature.

to 25 mL and cooled to 0° in a refrigerator overnight.* A crystalline material is deposited slowly that is removed by filtration. The mother liquor is cooled further to obtain two more crystalline crops at intervals of 24 hr.† The crude crops (2.7 g, mp 118–120°) are combined and recrystallized from petroleum at 0° (yield 2.5 g, 51.3%).

Anal. Calcd. for $C_{14}H_{41}N_{11}P_4$: C 34.5; H 8.5; N 31.6; P 25.4. Found: C 34.5; H 8.4; N 31.5; P 25.5.

Properties

The bicyclic phosphazene III forms colorless crystals that are stable in air and melt sharply at 124°. The compound is readily soluble in petroleum, benzene, chloroform, and dichloromethane but insoluble in water. The IR spectrum of the compound shows a broad absorption band at 1195 cm^{-1} attributable to ring P=N stretching vibration.[1,2] The ^{1}H NMR spectrum is discussed in the literature.[1,2] The ^{31}P {^{1}H} NMR spectrum is that of an A_2BC spin system (δ 18–24).[1] The structure of the compound has been established by X-ray crystallography.[8] The crystals are monoclinic with space group $C2/c$, $a = 10.81$, $b = 17.52$, $c = 28.48$ Å, $β = 94.8°$ with eight molecules in the unit cell.

C. *Trans*-2,4,4,6,8,8-HEXAKIS(DIMETHYLAMINO)-2,6-BIS(ETHYLAMINO)-1,3,5,7,2λ⁵,4λ⁵,6λ⁵,8λ⁵-TETRAAZATETRAPHOSPHOCINE (IV)

$$N_4P_4(NHEt)_2Cl_6 + 12HNMe_2 \xrightarrow{Et_2O}$$
II

$$N_4P_4 (NHEt)_2 (NMe_2)_6 + 6 HNMe_2 \cdot HCl$$
IV

Procedure

Compound II (1.92 g, 0.004 mol) (obtained by Procedure A) is dissolved in 250 mL of anhydrous diethyl ether and is treated with anhydrous dimethylamine (13.2 g, 0.3 mol) at 0° as described in Procedure B. The reaction mixture is stirred at 0° for 1 hr and allowed to attain ~25° in the course of 2 hr. Dimethylamine hydrochloride is removed by filtration.

---

*By rigorously excluding moisture during the work-up of the reaction mixture and by cooling the concentrated petroleum extract to 25° for 2 to 3 hr, it is possible to isolate a small quantity (<5%) of the hydrochloride adduct, $N_4P_4(NHEt)_2(NMe_2)_6 \cdot 2HCl$.[1]

†The remaining mother liquor contains mainly $N_4P_4(NHEt)_2(NMe_2)_6$ (IV) and traces of compound III as shown by TLC.[2,5]

Evaporation of the solvent from the filtrate yields a brownish oil that is extracted with petroleum (100 mL). The petroleum extract is washed with water to remove traces of dimethylamine hydrochloride. The organic layer is dried over anhydrous sodium sulfate, filtered and decolorized with activated charcoal (2 g). The clear filtrate is evaporated to dryness in a rotary evaporator to obtain 1.77 g (80%) of compound IV.

Anal. Calcd. for $C_{16}H_{48}N_{12}P_4$: C 36.1; H 9.1; N 31.6. Found: C 36.0; H 9.0; N 31.3.

Properties

Compound IV, isolated from the reaction mixture as a viscous liquid, solidifies on storage in a refrigerator for several months. Crystallization can be induced to a certain extent by occasionally scratching and thawing the cooled viscous liquid. The crystals of compound IV thus obtained melt between 136 to 138°. The compound is soluble in common organic solvents. The 1R spectrum of the compound shows a strong and broad absorption band centered at 1270 cm^{-1} attributable to ring P=N stretching vibration. The ^{1}H and proton decoupled ^{31}P NMR spectra (A_2B_2 spin system; $\delta_A = 9.2$, $\delta_B = 6.8$, $^2J_{AB} = 41.2$ Hz) are discussed in the literature.[5,11]

D. 9-ETHYL-1,3,3,5,7,7-HEXAKIS(ETHYLAMINO)-2,4,6,8,9-PENTAAZA-1λ⁵,3λ⁵,5λ⁵,7λ⁵-TETRAPHOSPHABICYCLO-[3.3.1] NONA-1,3,5,7-TETRAENE (V)

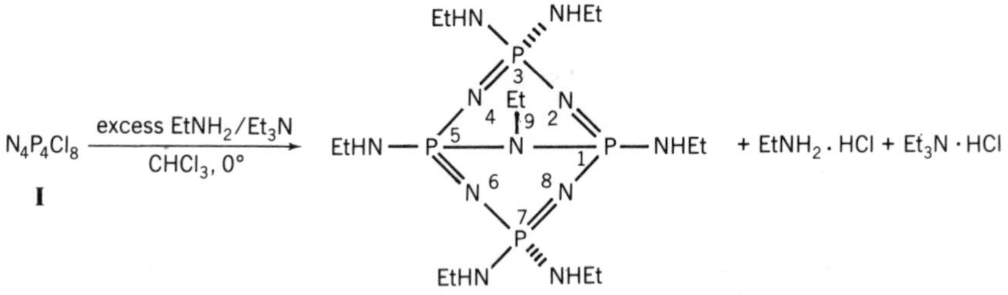

$$N_4P_4Cl_8 \xrightarrow[\text{CHCl}_3,\ 0°]{\text{excess EtNH}_2/\text{Et}_3\text{N}}$$

I

V

■ **CAUTION.** *Benzene is carcinogenic. An efficient hood should be used.*

Procedure

Compound I (9.28 g, 0.02 mol) is dissolved in 250 mL of chloroform (distilled over P_4O_{10}) in an apparatus similar to that used in Procedure B and the solution is cooled to 0°. Anhydrous ethylamine (26.2 g, 0.58 mol), previously cooled to ~ − 10°, is placed in the dropping funnel and is added to the reaction flask under vigorous stirring. An immediate reaction occurs as indicated by the formation of a curdy white precipitate of ethylamine hydrochloride. The stirring is continued at 0° for 4 hr and then the reaction mixture is allowed to attain ~25° during which time the excess of unreacted ethylamine escapes from the reaction flask. Ethylamine hydrochloride is removed by filtration and washed with 50 mL of chloroform. Evaporation of the solvent from the combined filtrate yields a residue consisting of a crystalline solid material dispersed in a viscous oil. The residue is extracted with hot benzene (200 mL), and the insoluble ethylamine hydrochloride is removed by filtration. Triethylamine (4.1 g, 0.04 mol) is added to the filtrate and the solution is heated under reflux for 2 hr. The solution is cooled and the precipitated triethylamine hydrochloride is removed by filtering. The filtrate is concentrated to 60 mL and cooled to ~25°. Crystals of crude product V are deposited slowly and are removed by filtration. The mother liquor is concentrated and cooled to obtain a further quantity of the crude product (total yield 4.7 g, 47%). The product is recrystallized from petroleum.

Anal. Calcd. for $C_{14}H_{41}N_{11}P_4$: C 34.5; H 8.5; N 31.6; P 25.4. Found: C 35.1; H 8.6; N 31.5; P 25.1.

Properties

The bicyclic phosphazene V forms colorless crystals, mp 184 to 185°. The compound is readily soluble in water, chloroform, and dichloromethane and is fairly soluble in hot benzene and moderately soluble in hot petroleum. The IR spectrum of the compound shows a split band at 1183 and 1218 cm^{-1}, attributable to a ring-P=N stretching vibration. The ^{1}H NMR spectrum is discussed in the literature.[1] The ^{31}P {^{1}H} NMR spectrum is that of an A_2B_2 spin system[1] (δ_A = 18.6, δ_B = 15.3, $^2J_{AB}$ = 40.9 Hz).

E. 2,6-BIS(*tert*-BUTYLAMINO)-2,4,4,6,8,8-HEXACHLORO- and 2,4-BIS(*tert*-BUTYLAMINO)-2,4,6,6,8,8-HEXACHLORO-1,3,5,7,2λ⁵,4λ⁵,6λ⁵,8λ⁵-TETRAAZATETRAPHOSPHOCINES (VI and VII)

$N_4P_4Cl_8$ + 4*t*-BuNH$_2$ $\longrightarrow$

   I

        2,6- and 2,4-$N_4P_4Cl_6$(NH-*t*-Bu)$_2$ + 2*t*-BuNH$_2$·HCl

                    VI, VII

Procedure

Compound I (9.28 g, 0.02 mol) is dissolved in 200 mL of acetonitrile (freshly distilled over P_4O_{10}) in a 500-mL, two-necked, round-bottomed flask, equipped with an ice–water cooled condenser, potassium hydroxide drying tube, and a magnetic stirring bar. The solution is heated to boiling and *tert*-butylamine* (5.84 g, 0.08 mol) in 20 mL of acetonitrile is added dropwise over a period of 1.5 hr. The reaction mixture is heated under reflux for 4 hr, cooled, and filtered to remove *tert*-butylamine hydrochloride. Evaporation of the solvent from the filtrate yields an oil that is extracted with 150 mL of petroleum (bp 60–80°). The petroleum solution is filtered to remove any insoluble material. The filtrate is concentrated (50 mL) and cooled to ~25°. The first two crops of crystals (mp 165–170°) isolated at intervals of 12 and 18 hr are combined and recrystallized from petroleum (50 mL) to obtain 2,6-bis(*tert*-butylamino)-hexachlorocyclotetraphosphazene (VI) (yield 3.5 g, 32.5%).† The solution, after removal of compound VI, is concentrated to 30 mL and cooled at 25°. Two crystalline crops of crude 2,4-isomer (VII) (mp 120–125°) are isolated at intervals of 12 hr (yield 1.5 g, 14%); recrystallization from petroleum affords a pure sample of compound VII,‡ mp 127°.

Anal. Calcd. for $C_8H_{20}N_6Cl_6P_4$: C 17.9; H 3.8; N 15.7; Cl 39.6. Found for compound VI: C 18.0; H 3.6; N 15.8; compound VII: C 17.8; H 3.6; N 15.9; Cl 39.8.

Properties

The two isomeric 2,6- and 2,4-bis(*tert*-butylamino) cyclotetraphosphazenes (VI, VII) melt at 171 and 127°, respectively. They are soluble in hot petroleum, benzene, and chloroform. Their IR spectra are similar.[12] The ring P=N vibration for the 2,6-isomer (VI) appears as a split band at 1300 cm^{-1}; for the 2,4-isomer (VII) this band is observed at 1310 cm^{-1}. The ^{31}P {^1H} NMR spectrum of the 2,6-isomer (VI) is of an A_2B_2 type, $\delta_A = -5.8$, $\delta_B = -10.6$, $^2J_{AB} = 38.1$ Hz.[13] The 2,4-isomer (VII) generates an $[AB]_2$-type ^{31}P NMR spectrum; $\delta_A = -8.7$, $\delta_B = -7.3$.[14] It is likely that in both the isomers, the amino groups have a trans orientation.[13,15]

*Dried over KOH pellets and distilled over sodium chips.
†The checkers report a yield of 21%.
‡The checkers note that separation of the 2,4-isomer (VII) from the reaction mixture is very difficult. They suggest separation of both isomers by high pressure liquid chromatography (HPLC) on a silica column using hexane–diethyl ether (20:1) as eluent. A better yield of 2,4-isomer (VII) can be obtained by carrying out the reaction in chloroform.[13]

F. 2,2,4,4,6,6,8,8-OCTAKIS(*tert*-BUTYLAMINO) 1,3,5,7,2λ⁵,4λ⁵,6λ⁵,8λ⁵-TETRAAZATETRAPHOSPHOCINE (VIII)

$$N_4P_4Cl_8 + 16t\text{-}BuNH_2 \longrightarrow N_4P_4 (NH\text{-}t\text{-}Bu)_8 + 8t\text{-}BuNH_2 \cdot HCl$$

I	VIII

Procedure

Compound I (4.64 g, 0.01 mol) is dissolved in 200 mL of acetonitrile in an apparatus similar to that used in Procedure E. The phosphazene solution is brought to boiling and *tert*-butylamine (17.0 g, 0.205 mol) in 50 mL acetonitrile is added dropwise over a period of 30 min. The reaction mixture is heated under reflux for 7.5 hr, cooled to ~25° and filtered. The filtrate is discarded and the precipitate is extracted with 100-mL petroleum using a Soxhlet apparatus (6 hr). The insoluble material is *tert*-butylamine hydrochloride. The petroleum extract is evaporated to yield a crystalline mass [mp 170–190° (d)]. The solid is dissolved in benzene–acetonitrile (1:1) (50 mL) and filtered. The filtrate is concentrated to 30 mL and cooled to 0° in a refrigerator for 24 hr. The crystals formed are removed by filtration. Further concentration and cooling of the mother liquor affords additional quantities of the desired product (VIII). The total yield of VIII is 3.9 g (52%).

Anal. Calcd. for $C_{32}H_{80}N_{12}P_4$: C 50.8; H 10.7; N 22.2; Found: C 50.5; H 10.7; N 22.6.

Properties

Compound VIII is a colorless solid [mp 180–200° (d)].* The compound is soluble in petroleum and benzene but is sparingly soluble in acetonitrile. The IR spectrum of the compound shows a strong band at 1235 cm⁻¹ attributable to ring-P=N stretching vibration.[12] The ^{31}P {^1H} NMR spectrum shows a single line at δ = −3.1.[13]

References

1. S. S. Krishnamurthy, A. C. Sau, A. R. Vasudeva Murthy, R. A. Shaw, M. Woods, and R. Keat, *J. Chem. Res.,* **1977,** (S) 70; (M) 0860–0884. (S = Synopsis version, M = Microfiche)

*The checkers report that the compound crystallizes as colorless needles from benzene containing 10 to 20% acetonitrile and it has a mp >230°.

2. S. S. Krishnamurthy, K. Ramachandran, A. C. Sau, R. A. Shaw, A. R. Vasudeva Murthy, and M. Woods, *Inorg. Chem.*, **18**, 2010 (1979).
3. S. S. Krishnamurthy, K. Ramachandran, and M. Woods, *J. Chem. Res.*, **1979**, (S) 92; (M) 1258–1266.
4. P. Ramabrahmam, K. S. Dhathathreyan, S. S. Krishnamurthy, and M. Woods, *Indian J. Chem.*, **22A**, 1 (1983).
5. S. S. Krishnamurthy, K. Ramachandran, and M. Woods, *Phosphorus Sulfur*, **9**, 323 (1981).
6. P. Y. Narayanaswamy, K. S. Dhathathreyan, and S. S. Krishnamurthy, *Inorg. Chem.*, **24**, 640 (1985).
7. P. Y. Narayanaswamy, S. Ganapathiappan, K. C. Kumara Swamy, and S. S. Krishnamurthy, *Phosphorus Sulfur*, **30**, 429 (1987).
8. T. S. Cameron and Kh.Mannan, *Acta Cryst.*, **33B**, 443 (1977); T. S. Cameron, R. E. Cordes, and F. A. Jackman, *Acta Cryst.*, **35B**, 980 (1979).
9. H. R. Allcock, *Polyhedron*, **6**, 119 (1987).
10. M. L. Nielson and G. Cranford, *Inorg. Synth.*, **6**, 94 (1960).
11. S. S. Krishnamurthy, A. C. Sau, A. R. Vasudeva Murthy, R. Keat, R. A. Shaw, and M. Woods, *J. Chem. Soc., Dalton Trans.*, **1976**, 1405.
12. S. S. Krishnamurthy, A. C. Sau, and M. Woods, *Advances in Inorganic and Radiochemistry*, Vol. 21, H. J. Emeléus and A. G. Sharpe (eds.), Academic Press, New York, 1978, p. 41.
13. S. S. Krishnamurthy, A. C. Sau, A. R. Vasudeva Murthy, R. Keat, R. A. Shaw, and M. Woods, *J. Chem. Soc., Dalton Trans.*, **1977**, 1980.
14. S. S. Krishnamurthy, K. Ramachandran, A. C. Sau, M. N. Sudheendra Rao, A. R. Vasudeva Murthy, R. Keat, and R. A. Shaw, *Phosphorus Sulfur*, **5**, 117 (1978).
15. S. S. Krishnamurthy, *Proc. Indian Natl. Sci., Acad.*, **52A**, 1020 (1986).

6. CHLOROFUNCTIONAL $1,3,5,2\lambda^5,4\lambda^5$-TRIAZADIPHOSPHININES

Submitted by ALFRED SCHMIDPETER*
Checked by PHILIP P. POWER† and RASIKA DAS†

Chlorofunctional triazadiphosphinines[1] range between 1,3,5-triazine and chloro-cyclotriphosphazenes (trimeric phosphonitrilic chlorides). They give substitution reactions[2-4] and thermal polymerization[1] like the latter. The tetrachloro derivatives may be prepared[1,5] in a [5 + 1] cyclocondensation from N,N-bis(trichlorophosphoranylidene)amidinium hexachloroantimonates and ammonium chloride or in a [3 + 3] cyclocondensation from ami-

*Institut für Anorganische Chemie, Universität München, D-8000 München 2, Federal Republic of Germany.
†Department of Chemistry, University of California, Davis, CA 95616.

dinium chlorides and μ-nitrido-bis(trichlorophosphorus)(1+) hexachloroantimonate:

$$Cl_3P \diagdown C(R) \diagup PCl_3^+ + NH_4^+ + 2Cl^- \diagdown$$

Partially organo substituted representatives can be prepared as well in analogy to the second route.

A.   μ-Nitrido-bis(trichlorophosphorus)(1+) Hexachloroantimonate[6]

$$SbCl_5 + HCl \longrightarrow HSbCl_6$$

$$NH_4 Cl + HSbCl_6 \longrightarrow [NH_4]SbCl_6 + HCl$$

$$[NH_4]SbCl_4 + 2PCl_5 \longrightarrow [N(PCl_3)_2]SbCl_6 + 4HCl$$

Procedure

In a 750-mL wide-necked Erlenmeyer flask 210 mL of concentrated hydrochloric acid are cooled in ice water. With magnetic stirring 300 g (1 mol) of antimony pentachloride (Aldrich, Fluka, Merck) are added from a dropping funnel within 1 hr (■ **Caution.** *Antimony pentachloride fumes in air. Avoid contact with skin and eyes and avoid breathing the corrosive vapors.*) Crystals of HSbCl₆·4.5H₂O separate from the solution. They are recrystalized from the mother liquor by gently warming (to 30–50°) and cooling to room temperature overnight. The crystals are separated on a coarse glass frit by suction. Yield: 322 g (77%) $HSbCl_6 \cdot 4.5H_2O$.[7],*

To a concentrated solution of 10.7 g (0.2 mol) of NH₄Cl in dilute hydrochloric acid a solution of 83.3 g (0.2 mol) of HSbCl₆·4.5H₂O in 20 mL of dilute hydrochloric acid is added with stirring. After 2 hr at 0° the

*Hexachloroantimonic acid may alternatively be prepared by adding chlorine to Sb₂O₃ suspended in concentrated hydrochloric acid.[8] For the precipitation of the ammonium hexachloroantimonate solution prepared in either way may be used directly without isolating the hexachloroantimonic acid hydrate.

colorless crystalline precipitate of [NH$_4$]SbCl$_6$ is separated on a glass frit, washed with cold hydrochloric acid, and dried at 80° in an oil pump vacuum. Yield: 38.5 g (55%).

In a dry 100-mL round-bottomed flask with inert gas inlet and reflux condenser topped by a calcium chloride tube 21.4 g (61 mmol) of [NH$_4$]SbCl$_6$ and 25.4 g (122 mmol) of PCl$_5$ are heated to reflux in 25 mL of nitromethane. While HCl evolves, all the solid material dissolves. The flask is flushed with dry nitrogen from time to time to expel the hydrogen chloride. After 2 to 3 hr the evolution of HCl should almost cease. On cooling the product separates as large crystals and is collected on a two-ended glass frit avoiding contact with moist air. After evaporating the mother liquor to half its volume a second crop is obtained. Yield: 31.6 g (84%). The crystals do not melt below 400°.

Anal. Calcd. for [Cl$_6$NP$_2$]SbCl$_6$ (623.1): Cl, 68.27; P 9.94. Found: Cl, 67.90; P 9.93.

B.　1,1,2-Trichloro-1-methyl-2,2-diphenyl-μ-nitridodiphosphorus(1+) Chloride

$$PhPSCl + 2NH_3 \longrightarrow PhPSNH_2 + NH_4Cl$$

$$MePCl_2 + Cl_2 \longrightarrow MePCl_4$$

$$Ph_2PSNH_2 + 2MePCl_4 \longrightarrow$$
$$[ClPh_2P{=}N{-}PMeCl_2]Cl + MePSCl_2 + 2HCl$$

Procedure

A 500-mL, three-necked, round-bottomed flask is equipped with stirrer, gas inlet, and paraffin oil-filled bubbler as outlet. Gaseous ammonia is passed into the stirred solution of 65.7 g (0.26 mol) of diphenylphosphinothioic chloride (prepared from commercially available Ph$_2$PCl and sulfur[9]) in 150 mL of diethyl ether for 2.5 hr. The precipitate forming consists of Ph$_2$PSNH$_2$ and NH$_4$Cl. It is separated on a funnel by suction and leached two times with 100 mL boiling benzene. On cooling Ph$_2$PSNH$_2$ separates from the filtrate in colorless crystals, mp 102–103°. Yield: 44.5 g (73%).*

In a 250-mL round-bottomed flask with inert gas inlet and equipped with reflux condenser and a calcium chloride tube 5.6 g (24 mmol) of

---

*Alternatively Ph$_2$PSNH$_2$ may be prepared from Ph$_2$PSCl and 20% aqueous ammonia.[10]

Ph$_2$PSNH$_2$ and 9.8 g (52 mmol) of MePCl$_4$ (freshly prepared from MePCl$_2$ (Merck, Strem) and Cl$_2$ in CCl$_4$)[11] in 100 mL of chloroform are slowly warmed to 60°. They react with rapid evolution of HCl. After removal of the chloroform in vacuum an oil is left. This is stirred with 50 mL of toluene, warmed to 60°, and thus caused to crystallize. The product is separated on a two-ended glass frit excluding moist air. Yield: 4.2 g (45%), moisture sensitive needles, mp 173–178° (with decomposition).

Anal. Calcd. for C$_{13}$H$_{13}$Cl$_4$NP$_2$ (387.0): C, 40.34; H, 3.38; N, 3.61. Found: C, 39.81; H, 3.47; N, 3.62.

^{31}P NMR(CH$_2$Cl$_2$): AB, δ_A = 42.6 (PPh), δ_B = 50.0 (PMe), J_{AB} = 28 Hz.

^{1}H NMR(CH$_2$Cl$_2$): δ_{CH_3} = 3.56 (dd, J_{PCH} = 16.0, J_{PNPCH} = 2.0 Hz).

C. 2,2,4,4-Tetrachloro-6-phenyl-1,3,5,2λ⁵,4λ⁵-triazadiphosphinine[1]

$$2[PhC(NH_2)_2]Cl + [N(PCl_3)_2]SbCl_6 \longrightarrow$$
$$PhC(NPCl_2)_2N + [PhC(NH_2)_2]SbCl_6 + 4HCl$$

Procedure

In a sublimation apparatus with a water-cooled probe 6.3 g (40 mmol) of carefully dried benzamidinium chloride[12,*] and 12.5 g (20 mmol) of [N(PCl$_3$)$_2$]SbCl$_6$ (see Procedure A) are thoroughly mixed and heated under a vacuum of 0.5 torr at 140 to 150° for 1 hr. The reaction mixture sinters and the product deposits at the probe as colorless crystals in 3.3 g (50%) yield, mp 90–93°.

Anal. Calcd. for C$_7$H$_5$Cl$_4$N$_3$P$_2$ (334.9): C, 25.11; H, 1.51; Cl, 42.35; N, 12.55; P, 18.50. Found: C, 25.06; H, 1.80; Cl, 42.53; N, 12.21; P, 19.09.

^{31}P NMR(CH$_2$Cl$_2$): δ = 41.6.

D. 2,2,4,4-Tetrachloro-6-(dimethylamino)-1,3,5,2λ⁵,4λ⁵-triazadiphosphinine[1]

$$2[Me_2NC(NH_2)_2]Cl + [N(PCl_3)_2]SbCl_6 \longrightarrow$$
$$Me_2NC(NPCl_2)_2N + [Me_2NC(NH_2)_2]SbCl_6 + 4HCl$$

---

*Commercially available [PhC(NH$_2$)$_2$]Cl·2H$_2$O (Aldrich, Fluka, Merck) loses the water at 100°.[13] It is used here after being kept at 120° and 0.5 torr for 1 hr.

Procedure

The 1,1-dimethylguanidinium chloride is prepared from the commercially available sulfate (Aldrich, Fluka): Sodium [9.2 g (0.4 mol)] is dissolved in 200 mL of methanol in a 500-mL round-bottomed flask with a reflux condenser. To the magnetically stirred solution 54.0 g (0.2 mol) of $[Me_2NC(NH_2)_2]_2SO_4$ are added. After 3 hr stirring the separating sodium sulfate is filtered off. Dry hydrogen chloride is passed into the filtrate until it is acidic and the methanol distilled off until $[Me_2NC(NH_2)_2]Cl$ starts to separate as colorless crystals. They are dried in vacuum at 60°. Yield: 29.8 g (60%). Addition of diethyl ether to the mother liquor gives a second crop.

In a 50-mL distilling flask* 3.9 g (32 mmol) of *N,N*-dimethylguanidinium chloride $[Me_2NC(NH_2)_2]Cl$ and 10.0 g (16 mmol) of $[N(PCl_3)_2]SbCl_6$ (see Procedure A) are heated at 120 to 140° under a vacuum of 0.2 torr. A melt is formed and within some minutes the product distills off and crystallizes in the receiving flask in colorless needles. Yield: 2.6 g (53%), mp 48–50°.

Anal. Calcd. for $C_3H_6Cl_4N_4P_2$ (301.9): C, 11.94; H, 2.00; N, 18.56. Found: C, 12.17; H, 1.99; N, 17.96.

^{31}P NMR(CH_2Cl_2): $\delta = 36.0$. ^1H NMR(CH_2Cl_2): $\delta_{CH_3} = 3.08$ (t, $J_{PNCNCH} = 0.6$ Hz).

E.   2,4-Dichloro-2,4,6-triphenyl-1,3,5,2λ^5,4λ^5-triazadiphosphinine[1]

$$[PhC(NH_2)_2]Cl + [N(PhPCl_2)_2]Cl \longrightarrow PhC(NPhPCl)_2N + 4HCl$$

Procedure

In a dry 100-mL round-bottomed flask closed by a calcium chloride tube 4.0 g (25.5 mmol) of benzamidinium chloride (see Procedure C) and 10.4 g (25.5 mmol) of $[N(PPhCl_2)_2]Cl^{14}$ are heated at 120 to 130° under a vacuum of 0.5 torr. The mixture melts and gives off hydrogen chloride. The evolution of HCl ceases after 0.5 hr. After cooling to room temperature the glassy product is dissolved in warm toluene and the solution is filtered. Removal of the solvent under vacuum leaves a colorless oil, which crystallizes within several hours. The product is a 1:2 mixture of the cis and trans isomers. Yield: 10.1 g (95%); mp 127–142°.

---

*The checkers used a short path distillation apparatus.

Anal. Calcd. for $C_{19}H_{15}Cl_2N_3P_2$ (418.2): C, 54.57; H, 3.62; N, 10.05. Found: C, 54.27; H, 3.22; N, 9.76.

^{31}P NMR(CH$_2$Cl$_2$): δ = 41.5 (presumably cis), 43.0 (presumably trans).

F. 2-Chloro-2-Methyl-4,4,6-triphenyl-1,3,5,2λ⁵,4λ⁵-triazadiphosphinine

$$[PhC(NH_2)_2]Cl + [ClPh_2P{=}N{-}PMeCl_2]Cl \longrightarrow$$
$$PhC(NPPh_2) \ (NPMeCl)N + 4HCl$$

Procedure

In a 50-mL round-bottomed flask closed by a calcium chloride tube 1.35 g (8.5 mmol) of benzamidinium chloride (see Procedure C) and 3.35 g (8.6 mmol) of [ClPh$_2$PNPMeCl$_2$]Cl (see Procedure B) are heated under a vacuum of 10^{-2} torr at 140 to 150° for 0.5 hr to give a clear melt. The glassy product is dissolved in 10-mL warm toluene and the solution is filtered through a two-ended glass frit. Removal of the solvent leaves a colorless crystalline substance. Yield: 3.00 g (88%); m.p. 90–96° (with dec).

Anal. Calcd. for $C_{20}H_{18}ClN_3P_2$ (397.8): C, 60.39; H, 4.56; N, 10.56. Found: C, 60.48; H, 4.98; N, 10.34.

^{31}P NMR(CH$_2$Cl$_2$): AB, δ$_A$ = 22.0 (PPh), δ$_B$ = 51.5 (PMe). ^{1}H NMR(CH$_2$Cl$_2$): δ$_{CH_3}$ = 2.06 (dd, J_{PCH} = 15.5, J_{PNPCH} = 1.7 Hz).

Properties

The chloro-triazadiphosphinines are readily soluble in a variety of solvents, such as dichloromethane or toluene. Compared to (NPCl$_2$)$_3$ they are more sensitive to moisture. At high temperature (~250°) they decompose, losing nitrile and forming a polymeric phosphazene. This decomposition may in certain cases be used[1] to initiate the polymerization of a cyclophosphazene. Nucleophilic substitution converts the chloro-triazadiphosphinines to partially or completely amino-, alkoxy-, or aroxysubstituted derivatives.[2–4]

References

1. A. Schmidpeter and N. Schindler, *Z. Anorg. Allg. Chem.*, **362**, 281 (1968).
2. A. Schmidpeter and N. Schindler, *Z. Anorg. Allg. Chem.*, **367**, 131 (1969).
3. A. Schmidpeter and N. Schindler, *Chem. Ber.*, **102**, 856 (1969).
4. A. Schmidpeter and N. Schindler, *Z. Anorg. Allg. Chem.*, **372**, 214 (1970).
5. A. Schmidpeter and R. Böhm, *Z. Anorg. Allg. Chem.*, **362**, 65 (1968).

6. A. Schmidpeter and K. Düll, *Chem. Ber.*, **100**, 1116 (1967).
7. E. Groschuff, *Z. Anorg. Allg. Chem.*, **103**, 164 (1918).
8. R. F. Weinland and H. Schmid, *Z. Anorg. Allg. Chem.*, **44**, 38 (1905).
9. N. K. Patel and H. J. Harwood, *J. Org. Chem.*, **32**, 2999 (1967).
10. I. N. Zhmurova, I. Yu. Voitsekhovskaya and A. V. Kirsanov, *Zh. Obshch. Khim.*, **29**, 2083 (1959); *J. Gen. Chem. USSR*, **29**, 2052 (1959); *Chem. Abs.* **54**, 8681 (1960).
11. R. Baumgärtner, W. Sawodny, and J. Goubeau, *Z. Anorg. Allg. Chem.*, **333**, 171 (1964).
12. A. W. Dox and F. C. Whitmore, *Org. Synth.* Coll. Vol. I, Second Edition, John Wiley, New York 1956, p. 5.
13. A. Pinner, Die Imidoäther und ihre Derivate, R. Oppenheim Verlag, Berlin 1892, S. 153.
14. E. Fluck and R. M. Reinisch, *Chem. Ber.*, **96**, 3085 (1963).

7. BINARY CYCLIC NITROGEN–SULFUR ANIONS

Submitted by J. BOJES,* T. CHIVERS,* and R. T. OAKLEY†
Checked by G. WOMERSHÄUSER‡ and M. SCHNAUBER‡

$1\lambda^4,3\lambda^4,5\lambda^4,7$-Tetrathia-2,4,6,8,9-pentaazabicyclo[3.3.1]nona-1(8),2,3,5-tetraenide anion, $N_5S_4^-$, was first prepared in 1975 by the methanolysis of $(Me_3E)N{=}S{=}N(SiMe_3)$ (E = C, Si).[1] It is also formed in the reaction of sulfur chlorides with anhydrous ammonia[2] and, in combination with $N_3S_3^-$, it is a frequent product of the nucleophilic degradation of N_4S_4.[3,4] It is conveniently prepared by the reaction of N_4S_4 with piperidine in absolute ethanol.[4] Oxidation of $N_5S_4^-$ with halogens, X_2, produces either N_5S_4Cl ($X_2 = Cl_2$) or N_6S_5 ($X_2 = Br_2$ or I_2).[5] The cage structure[6] of $N_5S_4^-$ breaks down on thermolysis to give the six-membered ring, $N_3S_3^-$, and subsequently, the acyclic NS_4^- ion.[7,8]

$1\lambda^4,3,5,2,4,6$-Trithiatriazenide anion, $N_3S_3^-$, has attracted considerable theoretical interest as a unique example of a 10 π-electron six-center cyclic system.[9–11] This anion is readily oxidized by molecular oxygen to give the oxyanions $N_3S_3O^-$ and $N_3S_3O_2^-$, in addition to $N_5S_4^-$ and $N_5S_4O^-$.[12] Salts of the type $M^+N_3S_3^-$ are best prepared by the reaction of N_4S_4 with ionic azides MN_3 [$M^+ = Cs^+$, R_4N^+, or $(Ph_3P)_2N^+$] in ethanol or acetonitrile.[3,7] With small alkali metal cations ($M^+ = Li^+$, Na^+, K^+) this procedure produces the $N_5S_4^-$ ion.[3]

*Department of Chemistry, The University of Calgary, Calgary, Alberta, Canada T2N 1N4.

†Department of Chemistry and Biochemistry, University of Guelph, Guelph, Ontario, Canada N1G 2W1.

‡Department of Chemistry, University of Kaisenslautern, Kaiserslautern D-6750. Federal Republic of Germany.

A. SALTS OF N₅S₄⁻ [1λ⁴, 3λ⁴, 5λ⁴,7-TETRATHIA-2,3,5,8,9-PENTAAZABICYCLO[3.3.1]NONA-1(8),2,3,5-TETRAENIDE]

$$4C_5H_{10}NH + 3N_4S_4 \longrightarrow$$

$$2[C_5H_{10}NH_2] [N_5S_4] + (C_5H_{10}N)_2S + \tfrac{3}{8}S_8 + N_2$$

1λ⁴,3λ⁴,5λ⁴,7-Tetrathia-2,4,6,8,9-pentaaza-bicyclo[3.3.1]nona-1(8),2,3,5-tetraenide

Procedure

■ **Caution.** *The recommended precautions for handling N₄S₄ are described in an earlier volume of Inorganic Syntheses.*[13] *These syntheses should not be attempted by anyone who lacks the experience or equipment (safety shields, goggles, and gloves) for handling these compounds. The reactions should not be carried out on a large scale. The alkali metal and ammonium salts of the N₅S₄⁻ ion are very susceptible to explosions under the influence of friction, pressure, or heat and the use of these counter ions should be avoided. The dangers of explosions can be minimized by the use of large cations, for example, Ph₄As⁺ or μ-nitrido-bis(triphenylphosphorus) (1+)(PPN⁺). Solid samples of these salts should not be removed from a glass frit with a metal spatula.*

Tetranitrogen tetrasulfide is prepared by the method of Villena-Blanco and Jolly.[14] Piperidine is stored over 4-Å molecular sieves and pentane over calcium hydride. Absolute ethanol is heated at reflux with magnesium turnings and a few crystals of iodine for 4 hr. The solvents and piperidine are distilled immediately before use. The reaction is conducted under an atmosphere of dry nitrogen in a one-necked flask (200 mL) equipped with a side arm.

Piperidine (1.86 g, 21.7 mmol) is added by syringe to a slurry of S₄N₄ (1.00 g, 5.43 mmol) in absolute ethanol (20 mL) and the solution is stirred (magnetic stirring bar) for ~4 hr at room temperature. Pentane (100 mL) is added to the bright orange solution to precipitate a flocculent yellow solid which, after filtration, is extracted (Soxhlet) with pentane on a fritted glass thimble for 18 hr to remove elemental sulfur. (■ **Caution.** *The solid product should not be removed from the glass frit with a metal spatula. Gentle tapping of the inverted glass frit, for example, with a rubber stopper* *attached to a glass rod, is recommended.*) The yellow product is then dissolved in absolute ethanol (10 mL) and is reprecipitated by the addition of pentane (60 mL). After filtration, analytically pure $[C_5H_{10}NH_2]$ $[N_5S_4]$ (0.60 g, 2.11 mmol) is obtained by washing with pentane (5 × 10 mL) and drying at 25° (10^{-2} torr) for 36 hr.

Anal. Calcd. for $C_5H_{12}N_6S_4$: C, 21.10; H, 4.26; N, 29.55; S, 45.08. Found: C, 21.07; H, 4.16; N, 29.47; S, 44.94.

μ-Nitrido-bis(triphenylphosphorus)(1+) $1\lambda^4,3\lambda^4,5\lambda^4,7$-tetrathia-2,4,6,8,9-pentaazabicyclo[3.3.1]nona-1(8),2,3,5-tetraenide, [PPN] $[N_5S_4]$ and $[Ph_4As]$ $[N_5S_4]$ are prepared in >90% yields by treatment of an aqueous solution of $[C_5H_{10}NH_2]$ $[N_5S_4]$ with equimolar amounts of [PPN]Cl (Aldrich) or $[Ph_4As]Cl$.

Properties

The compounds [PPN] $[N_5S_4]$ and $[Ph_4As]$ $[N_5S_4]$ are yellow, crystalline solids soluble in dichloromethane and warm acetonitrile or ethanol. Solid [PPN] $[N_5S_4]$ can be stored for years in a glass vial without special precautions. Solutions should be handled in an inert atmosphere to prevent oxidation or hydrolysis.

The characteristic IR absorptions of the $N_5S_4^-$ ion occur at ~950 (vs), 915 (vs), 740 (m), 720 (m), 690 (s), 655 (s), 600 (s), 530 (s), 505 (m), 440 (s) cm^{-1}.[1,3,4] The UV–vis spectrum of [PPN] $[N_5S_4]$ (in CH_2Cl_2) has a band at 293 nm (ϵ 5.9 × 10^3) and a shoulder at 345 nm (ϵ 2.5 × 10^3).[7] The ^{15}N NMR spectrum of [PPN] $[N_5^*S_4]$ (N* = 30% ^{15}N) in $CHCl_3$ shows singlets at +138.7(4N) and +53.3 ppm (1N) [ref. $NH_3(\ell)$].[15] The presence of the $N_3S_3^-$ ion as an impurity can be detected by the characteristic bands in the IR spectrum at ~640 cm^{-1} or in the UV–vis spectrum at ~365 cm (see below).

B. SALTS OF $N_3S_3^-$ [$1\lambda^4,3,5,2,4,6$-TRITHIATRIAZENIDE]

$$MN_3 + N_4S_4 \longrightarrow M[N_3S_3] + \tfrac{1}{8}S_8 + 2N_2$$

$1\lambda^4,3,5,2,4,6$-Trithiatriazenide

Procedure

■ **Caution.** *The recommended precautions for handling N_4S_4 are described in an earlier volume of Inorganic Syntheses.*[13] *These syntheses should not be attempted by anyone who lacks the experience or equipment (safety shields, goggles, and gloves) for handling these compounds. The reactions should not be carried out on a large scale. The alkali metal salts of the $N_3S_3^-$ ion may explode on grinding, scratching, or heating. It is recommended that the PPN^+ salt be used whenever possible. Solid samples of these salts should not be removed from a glass frit with a metal spatula.*

Tetramethylammonium azide is prepared by the neutralization of tetramethylammonium hydroxide (10% in water, Eastman) with a 10% solution of hydrazoic acid generated by passing an aqueous solution of sodium azide over Amberlite IR-120 ion-exchange resin.[16] The crude product is slurried twice with acetonitrile and, after filtration, the white solid residue is dissolved in the minimum amount of absolute ethanol and precipitated with diethyl ether. The white precipitate is dried at $60°/10^{-2}$ torr to give analytically pure [Me$_4$N] [N$_3$].[3] Tetranitrogen tetrasulfide is prepared by the method of Villena-Blanco and Jolly.[14] Absolute ethanol is heated at reflux with magnesium turnings and a few crystals of iodine for 4 hr and then distilled before use. The reaction and purification procedures are conducted under an atmosphere of dry nitrogen in a one-necked flask (200 mL) equipped with a side arm.

A slurry of N_4S_4 (0.84 g, 4.6 mmol) and [Me$_4$N] [N$_3$] (0.40 g, 3.4 mmol) in ethanol (40 mL) is stirred at room temperature (magnetic stirring bar). It is preferable to use a slight excess of N_4S_4 in order to avoid contamination of the product by unchanged azide. After 24 hr, the red solution, which contains a yellow precipitate of elemental sulfur and N_4S_4, is filtered and then treated with pentane (100 mL) to give a yellow solid. (■ **Caution.** *The solid product should not be removed from a glass frit with a metal spatula. Gentle tapping of the inverted glass frit, for example, with a rubber stopper attached to a glass rod, is recommended.*) This solid is removed by filtration and extracted (Soxhlet) with pentane on a fritted glass thimble (2 days) to remove sulfur and unreacted N_4S_4. The crude product is dissolved in absolute ethanol (20 mL). The solution is filtered and then treated with pentane (100 mL). The bright yellow precipitate is removed by filtration, washed with pentane (20 mL), and dried at $40°/10^{-2}$ torr for 24 hr to give [Me$_4$N] [N$_3$S$_3$] (0.39 g, 1.8 mmol).

Anal. Calcd. for C$_4$H$_{12}$N$_4$S$_3$: C, 22.61; H, 5.71; N, 26.38; S, 45.29. Found: C, 22.55; H, 5.59; N, 26.16; S, 45.23.

μ-Nitridobis(triphenylphosphorus)(1+)   1λ^4,3,5,2,4,6-trithiatriazenide is prepared from N_4S_4 and [PPN] $[N_3]^*$,[17] using a similar procedure. After extraction with pentane, the crude product is recrystallized from CH_2Cl_2–methanol to give lime-green blocks of [PPN] $[N_3S_3]$ in 66% yield.

Properties

Salts of the $N_3S_3^-$ ion are yellow or greenish-yellow solids, soluble in dichloromethane or acetonitrile and slowly oxidized on exposure to air. The PPN$^+$ salt is less susceptible to atmospheric oxidation than salts containing R_4N^+ or alkali metal cations. In solution, oxidation of $N_3S_3^-$ occurs more rapidly. It is easily detected by a change in color from yellow to red due to the formation of $N_3S_3O^-$ (λ_{max} 509 nm).[12] Solid samples should be stored at or below 0° in the absence of air and light to avoid oxidation and photochemical decomposition.

The characteristic IR absorptions of the $N_3S_3^-$ ion occur at ~920 (s), 640 (vs), and 380 (s) cm^{-1}.[9] The UV–vis spectra of $N_3S_3^-$ salts in CH_3CN exhibit a characteristic absorption at 365 nm (ϵ 8.2 × 10^3)[18] which has been attributed to the HOMO (highest occupied molecular orbital) $(2e'', \pi^*)$ → LUMO (lowest unoccupied molecular orbital) $(2a_2'', \pi^*)$ transition.[9] This assignment has been confirmed by the MCD (magnetic circular dichroism) spectrum of $N_3S_3^-$, which shows a negative A term for the 360-nm transition.[11] The ^{15}N NMR spectrum of 30% ^{15}N-enriched $N_3S_3^-$ as the PPN$^+$ salt in $CHCl_3$ shows a singlet at +148.4 ppm [ref. $NH_3(l)$].

References

1. O. J. Scherer and G. Wolmershäuser, *Angew. Chem. Int. Ed. (Engl.).*, **14**, 485 (1975).
2. O. J. Scherer and G. Wolmershäuser, *Chem. Ber.*, **110**, 3241 (1977).
3. J. Bojes and T. Chivers, *Inorg. Chem.*, **17**, 318 (1978).
4. J. Bojes, T. Chivers, I. Drummond, and G. MacLean, *Inorg. Chem.*, **17**, 3668 (1978).
5. T. Chivers and J. Proctor, *Can. J. Chem.*, **57**, 1286 (1978).
6. W. Flues, O. J. Scherer, J. Weiss, and G. Wolmershäuser, *Angew. Chem. Int. Ed. (Engl.)*, **15**, 379 (1976).
7. T. Chivers, W. G. Laidlaw, R. T. Oakley, and M. Trsic, *J. Am. Chem. Soc.*, **102**, 5773 (1980).
8. J. Bojes, T. Chivers, and R. T. Oakley, *Inorg. Synth.*, **25**, 37 (1988).
9. J. Bojes, T. Chivers, W. G. Laidlaw, and M. Trsic, *J. Am. Chem. Soc.*, **101**, 4517 (1979).
10. (a) T. Chivers, W. G. Laidlaw, and R. T. Oakley, *Inorg. Chim. Acta.*, **53**, L189 (1981). (b) V. H. Smith, Jr., J. R. Sabin, E. Broclawik, and J. Mrozek, *Inorg. Chim. Acta.*, **77**, L101 (1983). (c) M-T. Nguyen and T-K. Ha, *J. Mol. Struct.*, **105**, 129 (1983).
11. J. W. Waluk and J. Michl, *Inorg. Chem.*, **20**, 963 (1981).
12. T. Chivers, A. W. Cordes, R. T. Oakley, and W. T. Pennington, *Inorg. Chem.*, **22**, 2429 (1983).
13. A. J. Banister, *Inorg. Synth.*, **17**, 197 (1977).

*Available from Aldrich.

14. M. Villena-Blanco and W. L. Jolly, *Inorg. Synth.*, **9**, 98 (1967).
15. T. Chivers, R. T. Oakley, O. J. Scherer, and G. Wolmershäuser, *Inorg. Chem.*, **20**, 914 (1981).
16. V. Gutmann, G. Hampel, and O. Leitmann, *Monatsh. Chem.*, **95**, 1034 (1964).
17. J. K. Ruff and W. Schlientz, *Inorg. Synth.*, **15**, 84 (1974).
18. T. Chivers and M. Hojo, *Inorg. Chem.*, **23**, 1526 (1984).
19. Aldrichimica Acta, **16**, 84 (1983). Available from Aldrich.

8. BINARY CATENA-NITROGEN–SULFUR ANIONS

Submitted by J. BOJES,* T. CHIVERS,* and R. T. OAKLEY†
Checked by T. B. RAUCHFUSS‡ and S. GAMMON‡

The preparation of $[Bu_4N]$ $[NS_4]$ by the deprotonation of S_7NH with $[Bu_4N]$ [OH] has been described in an earlier volume of *Inorganic Syntheses.*[1] The thermolysis of the $N_3S_3^-$ ion, which need not be isolated, in acetonitrile represents an improved procedure for the preparation of salts of the NS_4^- ion.[2] The use of large cations, for example, μ-nitrido-bis-(triphenylphosphorus)(1+)(PPN$^+$) or Ph_4As^+, facilitates the isolation of crystalline salts suitable for X-ray structural determinations, which show that NS_4^- has an unusual structure in which nitrogen is the central atom of a planar, catenated chain with a cis,trans conformation.[2,3] The formation of the deep blue NS_4^- ion is dramatically apparent in the synthesis of cyclic sulfur imides from disulfur dichloride and ammonia[4] or sodium azide and sulfur[1] in polar solvents. This ion is also present in solutions of sulfur in liquid ammonia,[5] which have found application in organic syntheses.[6] The reaction of [PPN] $[NS_4]$ (or Ph_4As $[NS_4]$) with triphenylphosphine produces the corresponding salts of the NS_3^- ion,[7] which react with transition metal halides, for example, Cu(I), Ni(II), or Co(II), to give complexes of the type $M(NS_3)_2$.[7,8] The preparation of $Ni(NS_3)_2$ from N_4S_4 and nickel(II) chloride in methanol is reported in an earlier volume of *Inorganic Syntheses.*[9]

A. μ-NITRIDO-BIS(TRIPHENYLPHOSPHORUS)(1+) BIS(DISULFIDO)NITRATE(1−)

$$[PPN]N_3 + 2N_4S_4 \xrightarrow[\text{reflux}]{CH_3CN} [PPN] [NS_4] + \tfrac{1}{2}S_8 + 5N_2$$

*Department of Chemistry, The University of Calgary, Calgary, Alberta, Canada T2N 1N4.

†Departments of Chemistry and Biochemistry, University of Guelph, Guelph, Ontario, Canada N1G 2W1.

‡School of Chemical Sciences, University of Illinois, Urbana, IL 61801.

Procedure

■ **Caution.** *The cautionary notes that appear before the descriptions of the syntheses of* [PPN] [N$_5$S$_4$] *and* [PPN] [N$_3$S$_3$][10] *should be read before proceeding with these syntheses. Salts of the NS$_4^-$ or NS$_3^-$ ions do not undergo explosive thermal decomposition. They can be handled safely without special precautions except for the use of a dry inert atmosphere to prevent hydrolysis or oxidation.*

The starting materials and solvent are prepared and purified as described for the preparation of [PPN] [N$_3$S$_3$].[10] Toluene, pentane, and diethyl ether are dried over sodium, and the reaction is carried out in a 100-mL one-necked flask fitted with a side arm and under an atmosphere of dry nitrogen.

A slurry of [PPN]N$_3$ (1.50 g, 2.58 mmol)[11] and N$_4$S$_4$ (0.476 g, 2.58 mmol) in acetonitrile (45 mL) is stirred vigorously (magnetic stirring bar) for 1.5 hr to give a very dark blue-green solution, which is heated at reflux for a further 2 hr. The hot, royal blue solution is cooled to −20° for ~1 hr, filtered at room temperature to remove sulfur, and reduced in volume to ~15 mL. The solution is then kept at ~ −20° to complete the precipitation of [PPN] [NS$_4$] (~2 hr). The cold slurry is quickly filtered in air on a medium-porosity glass frit, and the purple microcrystalline product is washed with toluene (2 × 80 mL) and pentane (2 × 80 mL) to remove any remaining sulfur, and is stored under nitrogen. The yield of [PPN] [NS$_4$] is 1.35 g (77% based on S);* mp 155–158° (with decomposition).

Anal. Calcd. for C$_{36}$H$_{30}$N$_2$P$_2$S$_4$: C, 63.51; H, 4.44; N, 4.11; S, 18.84. Found: C, 63.37; H, 4.39; N, 3.98; S, 19.02.

Properties

The dark blue crystals of [PPN] [NS$_4$] have a copper-like metallic sheen. They are stable for several hours in air, but should be stored under nitrogen at 0°. The compound [PPN] [NS$_4$] is soluble in dichloromethane or acetonitrile and, in solution, the NS$_4^-$ ion is very sensitive to traces of oxygen and moisture decomposing to give oxoanions of sulfur. The UV–vis spectrum of [PPN] [NS$_4$] (in CH$_3$CN) exhibits an intense band at 582 nm (ε 16,000).[2] The characteristic vibrational frequencies of the NS$_4^-$ ion are IR (Nujol) 893 (w), 711 (m), 594 (s), 567 (vs), 416 (w); Raman (solid) 892 (w), 710 (vs), 592 (s), 570 (s) cm^{-1}.[2] The ^{15}N NMR spectrum of 30% ^{15}N-enriched NS$_4^-$ as the PPN$^+$ salt in CHCl$_3$ shows a singlet at +244.2 ppm [ref. NH$_3$(*l*)].[7]

---

*The checkers report a maximum yield of 51% even when the scale is increased by a factor of 2.

B. SALTS OF THE SULFIDO(DISULFIDO)NITRATE(1−) ANION, NS_3^-

$$M[NS_4] + Ph_3P \xrightarrow{CH_3CN} M[NS_3] + Ph_3PS$$

Procedure

Solid triphenylphosphine (0.728 g, 2.78 mmol) is added to a stirred royal blue solution of [PPN] [NS_4] (1.35 g, 1.98 mmol) in acetonitrile (30 mL) in a 100-mL one-necked flask fitted with a side arm and under an atmosphere of dry nitrogen. The optimum molar ratio of $Ph_3P:NS_4^-$ for maximizing the yield of NS_3^- is 1.4:1. Within 2 min the solution becomes orange. Slow addition of diethyl ether (~150 mL) precipitates a bright orange, microcrystalline solid, which is washed with acetonitrile–diethyl ether (1:5, 20 mL) and then diethyl ether (25 mL) to give analytically pure [PPN] [NS_3] (0.73 g, 1.13 mmol, 57%).*

Anal. Calcd. for $C_{36}H_{30}N_2P_2S_3$: C, 66.64; H, 4.67; N, 4.32; P, 9.55; S, 14.82. Found: C, 66.50; H, 4.52; N, 4.23; P, 10.06; S, 14.10.

If the orange crystals of [PPN] [NS_3] are redissolved in acetonitrile a blue solution containing NS_4^- is formed.[7] If recrystallization is necessary, acetonitrile *containing Ph₃PS* should be used as the solvent. The compound [Ph₄As] [NS_3] is prepared from [Ph₄As] [NS_4][3] and triphenylphosphine using a similar procedure.

Properties

Orange crystals of [PPN] [NS_3] can be handled in the air for 1–2 hr but darken on exposure to light for several days. Solid samples should be stored in the absence of air and light at 0°. The NS_3^- ion is readily converted to NS_4^- in solution (CH_3CN or CH_2Cl_2) or by the application of heat (>100°) or pressure (>1 ton) to solid samples. The UV–vis spectrum of [PPN] [NS_3], (in CH_3CN) exhibits an intense band at 465 nm accompanied by a weak absorption at 582 nm due to NS_4^-. The characteristic vibrational frequencies of the NS_3^- ion are IR(Nujol) 893 (vs), 666 (s), 574 (s); Raman (solid) 894 (m), 686 (s), 574 (vs) cm^{-1}.[7] The presence of small amounts of NS_4^- in solid samples of NS_3^- salts can be detected by the resonance Raman technique since the bands due to NS_4^- are greatly enhanced by use of a 600-nm exciting line.[7] The ^{15}N NMR spectrum of 30% ^{15}N-enriched NS_3^- as the PPN$^+$ salt in CHCl₃ shows a singlet at +617.6 ppm [ref. NH₃(*l*)].[7]

---

*The checkers report a 75% yield if the crystals of [PPN] [NS_3] are not washed with acetonitrile–diethyl ether.

References

1. J. Bojes, T. Chivers, and I. Drummond, *Inorg. Synth.*, **18**, 203 (1978).
2. T. Chivers, W. G. Laidlaw, R. T. Oakley, and M. Trsic, *J. Am. Chem. Soc.*, **102**, 5773 (1980).
3. N. Burford, T. Chivers, A. W. Cordes, R. T. Oakley, W. T. Pennington, and P. N. Swepston, *Inorg. Chem.*, **20**, 4430 (1981).
4. H. G. Heal and J. Kane, *Inorg. Synth.*, **11**, 184 (1968).
5. T. Chivers and C. Lau, *Inorg. Chem.*, **21**, 453 (1982).
6. R. Sato, T. Sato, K. Segawa, Y. Takikawa, S. Takizawa, and S. Oae, *Phosphorus Sulfur*, **7**, 217 (1979) and references cited therein.
7. J. Bojes, T. Chivers, W. G. Laidlaw, and M. Trsic, *J. Am. Chem. Soc.*, **104**, 4837 (1982).
8. J. Bojes, T. Chivers, and P. W. Codding, *J. Chem. Soc., Chem. Commun.*, **1981**, 1171.
9. D. T. Haworth, J. D. Brown, and Y. Chen, *Inorg. Synth.*, **18**, 124 (1978).
10. J. Bojes, T. Chivers, and R. T. Oakley, *Inorg. Synth.*, **25**, 30 (1988).
11. *Aldrichimica Acta*, **16**, 84 (1983). Available from Aldrich.

9. PENTANITROGEN TETRASULFIDE CHLORIDE, $[N_5S_4]Cl$ $[1\lambda^4,3\lambda^4,5\lambda^4,7$-TETRATHIA-2,4,6,8,9-PENTAAZABICYCLO[3.3.1]NONA-1(8),2,3,5-TETRAENYLIUM CHLORIDE]

$1\lambda^4,3\lambda^4,5\lambda^4,7$-Tetrathia-2,4,6,8,9-pentaaza-bicyclo[3.3.1]-1(8),2,3,5-tetraenylium

Submitted by T. CHIVERS* and L. FIELDING*
Checked by R. MEWS† and R. MAGGIULLI†

$$(NSCl)_3 + (Me_3Si)NSN(SiMe_3) \rightarrow [N_5S_4]Cl + 2Me_3SiCl$$

The preparations of the monocyclic S—N cations, $N_3S_4^+$ and $N_5S_5^+$, have been described in earlier volumes of *Inorganic Syntheses*.[1,2] The $N_5S_4^+$ ion is the only known example of a bicyclic, binary S—N cation. It is present in $[N_5S_4]Cl$, which has a polymeric chain structure with $N_5S_4^+$ cations bridged symmetrically by chloride ions.[3] The synthesis of $[N_5S_4]Cl$ is readily achieved by treatment of a carbon tetrachloride solution of trichlorocyclotrithiazene

*Department of Chemistry, The University of Calgary, Calgary, Alberta, Canada T2N 1N4.
†Institut Für Anorganische Chemie, Der Universität Göttingen, Tammanstrasse 4, D-3400 Göttingen, Federal Republic of Germany.

with bis(trimethylsilyl)sulfur diimide. It has been used to prepare $[N_5S_4]F,^{3,4}$ $[N_5S_4]$ $[AsF_6],^5$ and covalent derivatives of the N_5S_4 cage.[6]

Procedure

■ **Caution.** *Samples of* $[N_5S_4]Cl$ *may explode if subjected to friction, pressure, or sudden heating (e.g., during a melting point determination). This synthesis should not be attempted by anyone who lacks the experience or the necessary equipment (safety screen, goggles, and gloves) to handle this compound. The solid product should be handled with great care. It should not be removed from a glass frit with a metal spatula. The reaction should not be carried out on a large scale.*

1,3,5-Trichloro-1*H*,3*H*,5*H*-1,3,5,2,4,6-trithiatriazine is obtained by treatment of N_4S_4 with excess sulfonyl chloride.[2,7] Bis(trimethylsilyl)sulfur diimide is prepared by the reaction of $Na[N(SiMe_3)_2]$ with sulfinyl chloride by the literature method[8] with a few modifications as described by Kuyper and Street.[9] It is particularly important to wash the precipitate of sodium chloride thoroughly with diethyl ether to ensure of high yield of $(Me_3Si)NSN(SiMe_3)$. Carbon tetrachloride is stored over P_4O_{10} and distilled immediately before use. The reaction and work-up procedures are conducted in a 100-mL one-necked flask fitted with a side arm under an atmosphere of dry nitrogen.

Bis(trimethylsilyl)sulfur diimide (0.83 g, 4.03 mmol) is added dropwise* to a stirred solution (magnetic stirring bar) of $(NSCl)_3$ (0.95 g, 3.88 mmol) in carbon tetrachloride (50 mL) at 0°. The reaction mixture is allowed to stand at room temperature for ~15 hr, whereupon the supernatant liquid is removed by syringe from the crystalline product. If the solution of reagents is stirred during the reaction, the product is obtained as a yellow powder. Gold crystals are formed if the reaction mixture is not stirred. The product is washed with carbon tetrachloride (3 × 5 mL) and dried at 23°–10^{-2} torr for 2 hr to give analytically pure $[N_5S_4]Cl$ (0.56 g, 2.40 mmol).

Anal. Calcd. for ClN_5S_4: Cl, 15.17; N, 29.97; S, 54.86. Found: Cl, 15.23; N, 29.97; S, 54.58.

(■ **Caution.** *The dry solid product should not be removed from the flask or a glass frit with a metal spatula. Gentle tapping of the inverted flask or glass frit, for example, with a rubber stopper attached to a glass rod is recommended.*) Removal of solvent, under vacuum, from the combined washings and decanted solution produces additional $[N_5S_4]Cl$ (0.16 g, 0.69 mmol) as a yellow powder, which can be recrystallized from carbon tetrachloride. The total yield of $[N_5S_4]Cl$ is ~80%.

*The checkers preferred to add the $(Me_3SiN)_2S$ as a solution in 10 to 12 mL of CCl_4 in order to improve the accuracy of material transfer.

Properties

$1\lambda^4,3\lambda^4,5\lambda^4$-Tetrathia-2,4,6,8,9-pentazabicyclo[3.3.1]nona-1(8),2,3,5-tetra-enylium chloride is moisture sensitive. It is soluble in dichloromethane or liquid SO_2, slightly soluble in carbon tetrachloride and THF, and insoluble in hydrocarbon solvents. Solid samples should be stored in a dry atmosphere.

The characteristic IR absorptions of $[N_5S_4]Cl$ occur at 1140 (w), 1050 (s), 1023 (m), 966 (s), 800 (w), 720 (w), 697 (s), 628 (m), 584 (s) 560 (sh), 481 (m), 462 (m), 422 (m), 408 (m), 308 (w) cm^{-1}. The UV–vis spectrum (in CH_2Cl_2) exhibits a broad band at 255 mm (ϵ 1.5 × 10^4) with a shoulder at 365 mm (ϵ 2.3 × 10^3).

References

1. W. L. Jolly and K. D. Maguire, *Inorg. Synth.*, **9**, 102 (1967).
2. A. J. Banister and H. G. Clarke, *Inorg. Synth.*, **17**, 188 (1977).
3. T. Chivers, L. Fielding, W. G. Laidlaw, and M. Trsic, *Inorg. Chem.*, **18**, 3379 (1979).
4. W. Isenberg, R. Mews, G. M. Sheldrick, R. Bartetzko, and R. Gleiter, *Z. Naturforsch.*, **38b**, 1563 (1983).
5. W. Isenberg and R. Mews, *Z. Naturforsch.*, **37b**, 1388 (1982).
6. H. W. Roesky, C. Graf, and M. N. S. Rao, *Chem. Ber.*, **113**, 3815 (1980).
7. G. G. Alange, A. J. Banister, and B. Bell, *J. Chem. Soc., Dalton Trans.*, **1972**, 2399.
8. O. J. Scherer and R. Wies, *Z. Naturforsch.*, **25b**, 1486 (1970).
9. J. Kuyper and G. B. Street, *J. Am. Chem. Soc.*, **99**, 7848 (1977).

10. $(N_3P_2S)ClPh_4$, 1-Chloro-3,3,5,5-tetraphenyl-$1\lambda^4,2,4,6,3\lambda^5,5\lambda^5$-thiatriazadiphosphorine

1-Chloro-3,3,5,5-tetraphenyl-$1\lambda^4,2,4,6,3\lambda^5,5\lambda^5$-thiatriazadiphosphorine

Submitted by T. CHIVERS* and M. N. S. RAO†
Checked by R. T. OAKLEY‡

$$3N_4S_4 + 9Ph_2PCl \longrightarrow 3(N_3P_2S)ClPh_4 + 3Ph_2P(S)Cl + [N_3S_4]Cl + S_2Cl_2$$

1 - Chloro - 3,3,5,5 - tetraphenyl - $1\lambda^4,2,4,6,3\lambda^5,5\lambda^5$ - thiatriazadiphosphorine (N_3P_2S)ClPh$_4$, a six-membered heterocycle,[1,2] can be considered as a hybrid of the well-known cyclotriphosphazene, $(R_2PN)_3$,[3] and trichlorocyclotri-thiazene, $(NSCl)_3$,[4] ring systems. It is conveniently prepared by the reaction of diphenylphosphinous chloride with N_4S_4 in acetonitrile.[1] In addition to simple substitution reactions at the sulfur atom, $(N_3P_2S)ClPh_4$ readily participates in ring opening reactions via S—N bond cleavage to give 12-membered monocyclic or bicyclic compounds[5-7] and heterocycles containing a spirocyclic sulfur center.[1,7-9]

Procedure

■ **Caution.** *Tetranitrogen tetrasulfide may explode if it is subjected to percussion, friction, or sudden heating. Procedures for the safe handling of this material are given in ref. 10. This synthesis should not be attempted by anyone who lacks the experience or necessary equipment (safety screen, gloves, and goggles). The reaction of S_4N_4 with Ph_2PCl should not be carried out on a large scale.*

Tetranitrogen tetrasulfide is prepared by the method of Villena-Blanco and Jolly[11] and is recrystallized from toluene before use. Diphenylphosphinous chloride (Aldrich) is used as received. Acetonitrile is dried over phosphorus pentoxide and then calcium hydride, and distilled from CaH_2 when needed. The reaction is carried out in an oven-dried one-necked 100-mL flask fitted with a side arm and under an atmosphere of dry nitrogen, which is also used for the work-up procedure.

Diphenylphosphinous chloride (3.60 g, 16.3 mmol) in acetonitrile (10 mL) is added dropwise (15 min), via a dropping funnel, to a stirred suspension (magnetic stirrer) of tetranitrogen tetrasulfide (1.00 g, 5.4 mmol) in acetonitrile (25 mL). The solution is heated at reflux for 3 hr and the color of the solution changes from deep red to orange and, finally, yellow. The solution is allowed to cool to room temperature and then filtered to remove a yellow precipitate of [N_3S_4]Cl (0.45 g, 2.2 mmol). The filtrate is cooled to $-20°$ for 3–4 hr to give pale yellow crystals of (N_3P_2S)ClPh$_4$·CH$_3$CN (2.20 g, 4.2 mmol), which are collected by filtration at room temperature

*Department of Chemistry, The University of Calgary, Calgary, Alberta, Canada, T2N 1N4.

†*Present address:* Department of Chemistry, Indian Institute of Technology, Madras-36, India.

‡Department of Chemistry and Biochemistry, University of Guelph, Guelph, Ontario, Canada N1G 2W1.

and recrystallized from acetonitrile. The product so obtained contains 1 mol of solvated acetonitrile, which is removed on heating at $80°/10^{-2}$ torr to give $(N_3P_2S)ClPh_4$, mp 174-5°.

Anal. Calcd. for $C_{24}H_{20}ClN_3P_2S$: C, 60.06; H, 4.21; N, 8.76; P, 12.91. Found: C, 60.17; H, 4.29; N, 8.98; P, 12.59.

The unsolvated product is less sensitive to moisture then the acetonitrile adduct. The optimum yield of 75–80% (based on Ph_2PCl) is obtained for a 3:1 molar ratio of reactants ($Ph_2PCl:N_4S_4$). If this ratio is increased, significant amounts of the linear compound $[Ph_2P(Cl)N(Cl)PPh_2]Cl$ are formed.[1b,12]

Properties

1-Chloro-3,3,5,5-tetraphenyl-$1\lambda^4,2,4,6,3\lambda^5,5\lambda^5$-thiatriazadiphosphorine $(N_3P_2S)ClPh_4$, forms very pale yellow crystals that are readily hydrolyzed to give $[Ph_2P(NH_2)N(NH_2)PPh_2]^+Cl^-$.[13] It is slightly soluble in acetonitrile and readily dissolves in dichloromethane. In order to avoid hydrolysis, solvents must be scrupulously dried and solid samples of $(N_3P_2S)ClPh_4$ should not be stored in glass vessels for more than a few days.

The characteristic IR absorption bands of $(N_3P_2S)ClPh_4$ occur at 1439 (s), 1205 (vs), 1183 (s), 1160 (m), 1127 (vs), 1048 (s), 1000 (m), 728 (s), 692 (s), 548 (vs), 513 (s), 482 (m), 430 (m), and 385 (m) cm^{-1}. The purity of samples of $(N_3P_2S)ClPh_4$ is best checked by measurement of the ^{31}P NMR spectrum that consists of a singlet, $\delta = +7.8$ ppm (in $CDCl_3$, relative to external 85% H_3PO_4). The most likely impurity, the hydrolysis product (see above), is readily detected by a characteristic ^{31}P NMR signal at $\delta = +20.3$ ppm. The linear compound $[Ph_2P(Cl)N(Cl)PPh_2]Cl$ (in $CH_2Cl_2/CDCl_3$) exhibits a singlet at $\delta = +43.9$ ppm in the ^{31}P NMR spectrum.[1b]

References

1. (a) T. Chivers, M. N. S. Rao, and J. F. Richardson, *J. Chem. Soc., Chem. Commun.,* **1982**, 982; (b) T. Chivers and M. N. S. Rao, *Inorg. Chem.,* **23**, 3605 (1984).
2. N. Burford, T. Chivers, M. Hojo, W. G. Laidlaw, J. F. Richardson, and M. Trsic, *Inorg. Chem.,* **24**, 709 (1985).
3. R. Schmutzler, *Inorg. Synth.,* **9**, 75 (1967).
4. W. L. Jolly and K. D. Maguire, *Inorg. Synth.,* **9**, 102 (1967).
5. T. Chivers, M. N. S. Rao, and J. F. Richardson, *J. Chem. Soc., Chem. Commun.,* **1983**, 186.
6. T. Chivers, M. N. S. Rao, and J. F. Richardson, *J. Chem. Soc., Chem. Commun.,* **1983**, 702.

7. N. Burford, T. Chivers, M. N. S. Rao, and J. F. Richardson, *Adv. Chem. Ser.*, **232**, 81 (1983).
8. T. Chivers, M. N. S. Rao, and J. F. Richardson, *J. Chem. Soc., Chem. Commun.*, **1983**, 700.
9. T. Chivers, M. N. S. Rao, and J. F. Richardson, *Inorg. Chem.*, **24**, 2237 (1985).
10. A. J. Banister, *Inorg. Synth.* **17**, 197 (1977).
11. M. Villena-Blanco and W. L. Jolly, *Inorg. Synth.*, **9**, 98 (1967).
12. W. Haubold, D. Kammel, and M. Becke-Goehring, *Z. Anorg. Allg. Chem.*, **380**, 23 (1971).
13. I. I. Bezman and J. H. Smalley, *Chem. Ind. (London)*, **1960**, 839.

11. SILICON AND TIN SULFUR–NITROGEN COMPOUNDS, $(Me_3Si)_2N_2S$, $(Me_3Sn)_2N_2S$, and $(Me_2Sn)S_2N_2$

Submitted by CHRISTOPHER P. WARRENS and J. DEREK WOOLLINS*
Checked by M. WITT and H. W. ROESKY†

Bis(trimethylsilyl)sulfurdiimide has been known for a number of years.[1] It may be obtained from reaction of $Na[(Me_3Si)_2]$ with sulfur dichloride or, in better yield, sulfinyl chloride.[2] Its synthetic utility has been thoroughly established;[3] elimination of Me_3SiCl with insertion of the NSN unit provides many useful syntheses. Its tin analog is also known[4] and of similar usefulness. The compound $(Me_3Sn)_2N_2S$ has been obtained[4,5] from the reaction of $(Me_3Sn)_3N$ or Me_3SnNMe_2 with S_4N_4. An alternative reaction[6] that does not require the use of explosive S_4N_4 and is easier to work-up is that between $(Me_3Sn)_3N$ and $(NSCl)_3$. The S → Sn coordination dimer, $[(Me_2Sn)S_2N_2]_2$, obtained[4] from the reaction of S_4N_4 with $(Me_3Sn)_3N$, has also found use in the synthesis of SN heterocycles. Despite the utility of these compounds, well documented and foolproof procedures have not been reported and this is limiting the development of their chemistry. In the following sections we provide detailed revised procedures. Each preparation takes ~2 days and may be scaled up or down by a factor of 2 although the preparation of $(Me_3Si)_2N_2S$ does become cumbersome and the mixture is difficult to stir if scaled up because of the large quantities of solvent needed.

*School of Chemical Sciences, University of East Anglia, Norwich, NR4 7TJ, United Kingdom. (Work carried out at Imperial College of Science and Technology, London, United Kingdom.

†Institut für Anorganische Chemie, Der Universitat Göttingen, Tammannstrasse 4, D-3400 Göttingen, Federal Republic of Germany.

Procedure

All procedures should be carried out under an inert atmosphere of nitrogen or argon. Solvents used should be anhydrous. Diethyl ether is dried over sodium, dichloromethane is dried with and distilled from CaH_2.

A. BIS(TRIMETHYLSILYL)SULFURDIIMIDE $(Me_3Si)_2N_2S$

$$2(Me_3Si)_2NNa + SOCl_2 \longrightarrow (Me_3Si)_2N_2S + (Me_3Si)_2O + 2NaCl$$

Sodium bis(trimethylsilyl)amide[7] (99.06 g, 0.54 mol) (checkers report that the lithium salt gives slightly better yields) is dissolved in diethyl ether (550 mL) in a 1-L round-bottomed flask equipped with a nitrogen inlet, a 50-mL pressure equalizing dropping funnel, and a Teflon covered magnetic stirring bar (if the reaction is to be carried out on a larger scale a mechanical stirrer should be used). The reaction vessel is cooled using a CO_2–acetone bath ($-78°$). Sulfinyl chloride (33 g, 20.2 mL, 0.277 mol) is added to the stirred solution via the dropping funnel over 30–45 min and with continued stirring the reaction mixture is allowed to warm slowly to room temperature (3–4 hr). At this stage the mixture is yellow–pale orange and contains large quantities of precipitated NaCl, which is removed by filtration through a Schlenk filter (7.5-cm diameter, medium porosity, D #3 frit). The NaCl is thoroughly washed with diethyl ether to remove any product adhering to it. Typically this requires 4×150 mL and washing should be continued until the NaCl is almost white. The filtrate and washings are combined and the diethyl ether removed by distillation at atmospheric pressure under nitrogen. Some further NaCl may be precipitated on removal of the ether and this should be removed by filtration. The product is obtained by fractional distillation via a Vigreux column (30 cm $\times$ 1 cm) at reduced pressure (12 torr—a water pump may be used provided the distillation is protected by a drying tube) as a pale yellow mobile liquid, bp 59–61°, 12 mm; yield: 33.4 g, 60%.

Anal. Calc. for $[(CH_3)_3Si]_2N_2S$: C 34.91, H 8.79, N 13.57, S 15.53%. Found: C 34.74, H 8.95, N 13.36, S 15.61%.

Properties

The compound $(Me_3Si)_2N_2S$ is a pale yellow, moisture sensitive liquid. It should be stored under nitrogen, preferably in a flask equipped with Teflon in glass valves (e.g., Rotaflo or J. F. Youngs type) since it slowly dissolves grease. It is miscible with, and may be used in, dry organic solvents such as hexane, benzene, and THF, but it reacts with alcohol.[8] The ^{1}H NMR (CDCl$_3$) consists of a singlet at δ = 0.26 ppm IR (cm^{-1}) (thin film) 2970 (m), 2910 (w), 1250 (s, br), 1145 (w), 1062 (w), 942 (w), 850 (s, br), 770 (m), 725 (mw), 705 (sh), 658 (w), 628 (w), 455 (mw). Mass Spectroscopy: [m/e(rel intensity)—major peaks only] 206 (12) M$^+$, 191 (100), 177 (20), 146 (45), 138 (25), 120 (30), 73 (95).

B. BIS(TRIMETHYLSTANNYL)SULFURDIIMIDE (Me$_3$Sn)$_2$N$_2$S

$$(Me_3Sn)_3N + \tfrac{1}{3}(NSCl)_3 \longrightarrow (Me_3Sn)_2N_2S + Me_3SnCl$$

■ **Caution.** *The product obtained in this reaction is malodorous and of unknown toxicity. It should be handled in an efficient fume hood.*

Tris(trimethylstannyl)amine[9] (5.225 g, 0.0103 mol) dissolved in (sodium dried) toluene (50 mL) is placed in a 500-mL Schlenk flask equipped with a 250-mL pressure equalizing dropping funnel and a Teflon coated magnetic stirring bar. The dropping funnel is charged with a warm (50°) solution of trithiazltrichloride[10] (0.8484 g, 0.00347 mol) in dry toluene (150 mL). Over a period of 1 to 2 hr this solution, maintained at ~50° (using a hot air blower) to stop precipitation of (NSCl)$_3$, is added dropwise to the stirred reaction mixture. The reaction mixture is stirred overnight and the toluene evaporated under reduced pressure (15 torr) with no external heating to leave an orange residue. This residue is fractionally sublimed using a cold finger cooled with liquid N$_2$. At room temperature and 0.05 torr Me$_3$SnCl is obtained while the product sublimes at 0.05 torr and 50–60°. Yield: 2.091 g, 52%. (Checkers comment: alternatively, Me$_3$SnCl can be removed by trapping it in liquid N$_2$ during sublimation of the product onto a water cooled sublimation finger. On a larger scale pumping for 1 hr prior to sublimation is advisable.)

Anal. Calc. for [(CH$_3$)$_3$Sn]$_2$N$_2$S: C 18.59, H 4.68, N 7.23, S 8.27%. Found: C 18.63, H 4.69, N 7.05, S 8.10%.

Properties

The compound (Me$_3$Sn)$_2$N$_2$S is a pale yellow air- and moisture-sensitive, malodorous solid that must be stored under nitrogen. If kept at 0° it can be stored indefinitely. It is soluble in most organic solvents (e.g., diethyl ether, benzene, THF, CH$_2$Cl$_2$). The ^{1}H NMR (CDCl$_3$) consists of a singlet at δ = 0.43 ppm with tin satellites [${}^2J_{\text{Sn–H}}$ 57 Hz], whereas in the ^{119}Sn {^{1}H}

NMR (C_6D_6) a broad singlet at $\delta = 40.5$ ppm (rel. to Me_4Sn) is observed. Infrared (Nujol mull), 1200 (vs), 1170 (vs), 1057 (w), 1000 (s, br), 933 (m, br), 770 (vs, br), 645 (s), 540 (s), 522 (s), 509 (s), 403 (s, br). The mass spectrum shows the parent ion at m/e 388 together with the reported fragmentation products.[6]

C. BIS[MERCAPTOSULFURDIIMIDATO(2−)] TETRAMETHYLDITIN, [Me₂SnS₂N₂]₂

$$2(Me_3Sn)_3N + S_4N_4 \longrightarrow 2[(Me_3Sn)_2N_2S] + Me_4Sn + \tfrac{1}{2}$$

(■ **Caution.** *Tetranitrogen tetrasulfide used in this procedure is explosive. The greatest of care should be taken in handling S_4N_4.*) Under no circumstances should it be heated suddenly or struck; care should be taken to avoid trapping it in ground glass joints. Refer to the previous safety note.[11] The reaction should be carried out in a fume hood since some of the tin compounds formed have unpleasant odors and are of unknown toxicity.

Tetranitrogen tetrasulfide[12] (1.0 g, 0.0054 mol) is suspended in dry CH_2Cl_2 (60 mL) in a 100-mL Schlenk flask (i.e., a 100-mL round-bottomed flask with a single quickfit joint and a stopcock) equipped with a 50-mL pressure equalizing dropping funnel and a Teflon coated magnetic stirrer bar. This suspension is cooled to $-20°$ and a solution of tris(trimethylstannyl)amine[9] (5.47 g, 0.0108 mol) in CH_2Cl_2 (18 mL) is slowly added dropwise over 1 hr to the stirred suspension. The reaction mixture is allowed to warm to room temperature ($\sim$1 hr) and the solvent is removed under reduced pressure with no external heating of the sample to give a sticky orange colored residue. A second 100-mL Schlenk flask is connected to the reaction vessel via an adapter bend and is immersed in a bath at $-40°$. The receiver flask is connected via a liquid N_2 cooled cold trap to a vacuum line, and at a pressure of 0.01 torr, the reaction flask is heated to 70°. Tetramethyltin (Me_4Sn) is collected in the liquid N_2 cooled trap, while impure $(Me_3Sn)_2N_2S$ distills over into the receiver flask ($-40°$) leaving a yellow residue in the reaction flask. The yellow residue is recrystallized from boiling CH_2Cl_2 (~100 mL) to give the product as a yellow powder. The impure $(Me_3Sn)_2N_2S$, which is contaminated with a more volatile compound, may be purified by distillation (15 × 1 cm Vigreux column) at 0.175 mm Hg. The impurity distils at 66 to 68° leaving behind a residue that can be sublimed as described above to give pure $(Me_3Sn)_2N_2S$, 1.95 g, 46%. Yield: $(Me_2Sn)S_2N_2$ 0.572 g, 44%.

Anal. Calc. for $(CH_3)_2SnS_2N_2$: C 9.97, H 2.51, N 11.63, S 26.62%. Found: C 10.30, H 2.51, N 10.98, S 26.88%.

The checkers prefer a slightly modified procedure. After addition of the amine the solution is refluxed for several hours, the sulfurdiimide is sublimed off at 50–70° *in vacuo* and the residue washed with several portions of CCl_4. The product thus obtained is sufficiently pure for most purposes.

Properties

The compound $[Me_2SnS_2N_2]_2$ is a slightly air sensitive yellow solid, which is moderately soluble in $CHCl_3$, CH_2Cl_2, toluene, benzene, and DMF. The ^1H NMR ($CDCl_3$) consists of a singlet at $\delta = 0.934$ ppm with tin satellites, $^2J_{Sn-H} = 66$ Hz. Infrared (Nujol mull) 2980 (w), 2900 (w), 1186 (m), 1064 (s, br), 1034 (vs, br), 900 (s), 774 (s, br), 703 (s), 626 (s), 566 (m), 526 (m), 458 (m), 396 (m). Mass spectral measurements show the parent ion of the monomer at m/e 242 and peaks due to the expected fragmentation products.

References

1. U. Wannagat and H. Kuckertz, *Angew. Chem. Int. Ed. (Engl.)*, **1**, 113 (1962).
2. O. J. Scherer and R. Wies, *Z. Naturforsch.*, **25B**, 1486 (1970).
3. H. W. Roesky, *Adv. Inorg. Radiochem.*, **22**, 239 (1979).
4. H. W. Roesky and H. Wiezer, *Angew. Chem. Int. Ed. (Engl.)*, **12**, 674 (1973).
5. D. Hanssgen and W. Roelle, *J. Organomet. Chem.*, **56**, C14 (1973).
6. G. Brands and A. Golloch, *Z. Naturforsch.*, **37B**, 568 (1982).
7. C. R. Kruger and H. Niederprum, *Inorg. Synth.*, **8**, 15 (1966).
8. W. Flues, O. J. Scherer, J. Weiss, and G. Wolmershauser, *Angew. Chem. Int. Ed. (Eng.)*, **15**, 379 (1976).
9. W. L. Lehn, *J. Am. Chem. Soc.*, **86**, 305 (1964).
10. W. L. Jolly and K. Maguire, *Inorg. Synth.*, **9**, 102 (1967).
11. A. J. Banister, *Inorg. Synth.*, **17**, 197, (1977).
12. M. Villena-Blanco and W. L. Jolly, *Inorg. Synth.*, **9**, 98 (1967).

12.   1,1,1-TRIMETHYL-*N*-SULFINYLSILANAMINE, Me₃SiNSO

Submitted by ERICA PARKES and J. DEREK WOOLLINS*
Checked by HOWARD B. YOKELSON† and ROBERT WEST†

1,1,1-Trimethyl-*N*-sulfinylsilanamine, Me₃SiNSO, obtained from the re-action of (Me₃Si)₃N with SOCl₂ was first reported[1] in 1966 but full details of the preparation have not been published. This compound provides a useful source of the NSO function by reactions involving elimination of Me₃SiCl, for example, in the formation[2] of S(NSO)₂, and it may be con-verted into a stable mercury salt Hg(NSO)₂ by reaction[3] with HgF₂.

The procedure described here can be carried out comfortably in 2 days and may be scaled up twofold provided the reaction time is increased accordingly.

Procedure

All manipulations are performed under an inert atmosphere of nitrogen or argon. Aluminum trichloride is sublimed under vacuum (120°) onto a dry ice–acetone cold finger and should be white in color. Sulfinyl chloride was distilled from sulfur or is freshly opened Aldrich Gold Label.

$$(Me_3Si)_3N + SOCl_2 \xrightarrow{AlCl_3} 2Me_3SiCl + Me_3SiNSO$$

1,1,1-Trimethyl-*N*,*N*-bis(trimethylsilyl)silanamine[4] (20.0 g, 0.086 mol), aluminum trichloride (0.5 g) and a magnetic stirring bar are placed in a two-necked, 100-mL, round-bottomed flask equipped with a pressure equalizing dropping funnel containing sulfinyl chloride (10.23 g, 0.086 mol) and a reflux condenser, which is fitted with an N₂ bubbler. The sulfinyl chloride is added dropwise from the dropping funnel with stirring over a period of 1 hr during which time the reaction mixture changes from white to orange-yellow. The reaction mixture is then stirred and heated to 70° (oil bath) for 24 hr. After this time the reaction is cooled. The dropping funnel and condenser are removed and quickly replaced by a distillation head (a Perkin triangle is most convenient but a simple arrangement with

---

*School of Chemical Sciences, University of East Anglia, Norwich, NR4 7TJ, United Kingdom.

†Department of Chemistry, University of Wisconsin, Madison, WI 53706.

a "pig" is satisfactory). Distillation under N_2 gives two low boiling fractions (Me_3SiCl and $SOCl_2$) followed by the pure product, bp 102–4°, as a yellow liquid. Yield: 5.9 g, 51%. (Checkers report 40% yield for a reaction çarried out on half-scale).

Anal. Calc. for $(CH_3)_3SiNSO$: C·26.66 H 6.66 N 10.37%. Found: C 26.87 H 6.85 N 10.17%.

Properties

1,1,1-Trimethyl-N-sulfinylsilanamine is a slightly air–moisture sensitive yellow liquid and is best stored under nitrogen in a greaseless Schlenck flask. It is soluble in most organic solvents but decomposes in alcohols. ^1H NMR (CD_2Cl_2) singlet $\delta = 0.35$ ppm. Infrared (thin film, cm^{-1}) 2960 (m), 2900 (w), 1405 (w), 1295 (vs, br) (NSO), 1125 (s) (NSO), 1050 (w), 845 (vs, br), 760 (s), 695 (w), 640 (s), 570 (m) (NSO), 465 (w).

References

1. O. J. Scherer and P. Hornig, *Angew. Chem. Int. Ed. (Engl.),* **5,** 729 (1966).
2. D. A. Armitage and A. W. Sinden, *Inorg. Chem.,* **11,** 1151 (1972).
3. W. Verbeek and W. Sundermeyer, *Angew. Chem. Int. Ed. (Engl.),* **8,** 376 (1969).
4. W. L. Lehn, *J. Am. Chem. Soc.,* **86,** 305 (1964).

13. SULFUR–NITROGEN RINGS CONTAINING EXOCYCLIC OXYGEN

Submitted by HERBERT W. ROESKY* and MICHAEL WITT*
Checked by JOHN BURGESS,† PAUL F. KELLY,† THOMAS G. PURCELL,† and J. DEREK WOOLLINS†

While tetrasulfurtetranitride[1] and sulfur–nitrogen halides[2] have been established over a century ago, progress in the field of oxygen-containing ring compounds has not been achieved until the last decade. Almost si-

*Institute für Anorganische Chemie, der Universitat Göttingen, Tammännstrasse 4, D-3400 Göttingen, Federal Republic of Germany.
†School of Chemical Science, University of East Anglia, Norwich, NR4 7TJ, United Kingdom.

multaneously an eight-membered ring[3] and a five-membered ring[4] were synthesized [see eqs. (1) and (2)].[5]

$$4S_3N_2Cl_2 + 2SO_2(NH_2)_2 \longrightarrow$$
$$2S_4N_4O_2 + S_4N_3Cl + NH_4Cl + S_2Cl_2 + 4HCl \quad (1)$$
$$S_3N_2Cl_2 + HCOOH \longrightarrow S_3N_2O + 2HCl + CO \quad (2)$$

The synthesis of these two rings, together with the first synthesis of a carbon-containing five-membered ring, are given in detail [see eqs. (3)[6] and (4)[7]].

$$S_2N_2Sn(CH_3)_2 + COF_2 \longrightarrow S_2N_2CO + 2(CH_3)_2SnF_2 \quad (3)$$
$$ClCOSCl + (CH_3)_3SiNSNSi(CH_3)_3 \longrightarrow S_2N_2CO + 2(CH_3)_3SiCl \quad (4)$$

While the eight-membered ring proved to be an excellent precursor for the synthesis of oxygen-containing sulfur–nitrogen heterocycles with nucleophiles,[8] as well as electrophiles,[9] the five-membered rings have shown Lewis basicity on various substrates.[10,11]

General Procedure

All reactions were carried out in a well-ventilated hood in a dry nitrogen atmosphere. All solvents were dried according to the literature and were stored under nitrogen. The compounds $SO_2(NH_2)_2$ and $(CH_3)_3SnCl$ were commercial products (e.g., Morton Thiokol Inc., Alfa Products), COF_2 can be purchased, for example, from PCR Inc., Gainesville, Florida, HCOOH was dried with boron oxide/anhydrous copper(II)chloride, $S_3N_2Cl_2$[12] S_4N_4,[13] and $[(CH_3)_3Sn]$[14] were prepared according to the literature methods.
   ■ **Caution.** *S_4N_4 should be handled with extreme caution and in small quantities. It is prone to detonate.*[13d]

A. TETRASULFURTETRANITROGEN DIOXIDE, $S_4N_4O_2$

Tetrasulfurtetranitrogen dioxide is best prepared according to the following equation

$$2SO_2(NH_2)_2 + 4S_3N_2Cl_2 \longrightarrow$$
$$2S_4N_4O_2 + S_4N_3Cl + NH_4Cl + S_2Cl_2 + 4HCl$$

500 mL of dry carbon tetrachloride are heated to reflux in a 1-L three-necked flask equipped with a reflux condenser topped with a $CaCl_2$ drying tube, a solid addition funnel charged with a mixture of 78 g finely ground $S_3N_2Cl_2$ (0.4 mol) and 19.2 g (0.2 mol) $SO_2(NH_2)_2$* and a mechanical stirrer or a strong magnetic stirrer. Through a gas inlet on top of the funnel a slow stream of nitrogen is maintained throughout the whole reaction to avoid blocking of the solid outlet and to expel HCl. The mixture is added to the boiling solvent over a period of 3 hr and the slurry is heated until HCl evolution has ceased (approximately 8 hr). After cooling, the dark yellow precipitate is collected under nitrogen on a sintered glass funnel and dried *in vacuo*. The filtrate can be discarded; it contains ~1 to 2% of $S_4N_4O_2$. Extraction of the solid with dry, peroxide-free THF, using a soxhlet apparatus with a sintered glass extraction thimble, yields pure $S_4N_4O_2$ after cooling. More product can be obtained by concentrating the solution. The overall yield is strongly dependent on the reaction scale; smaller scales decrease the yield drastically.[15] Thus an optimized yield up to 90% can sometimes be achieved. (The checkers obtained 56–60% yields. In a one-tenth scale synthesis, the yield fell to 37%.)

Properties

Tetrasulfurtetranitrogen dioxide $(1\lambda^6,3\lambda^4,5,7\lambda^4$-tetrathia-2,4,6,8-tetraazocine$)^5$ forms orange crystals melting with decomposition at 168°. It crystallizes in the monoclinic space group $P2_1/c$,[16] five ring atoms being essentially planar, with the two nitrogen atoms adjacent to the SO_2 moiety 55 pm and the tetra co-ordinated sulfur 157 pm out of plane. The compound $S_4N_4O_2$ is soluble in benzene, acetronitrile, THF, and liquid sulfur dioxide, slightly soluble in methylene chloride, and almost insoluble in petroleum and chlorofluorocarbons. Pyridine forms an adduct. In water and alcohols decomposition takes place.[8b]

The IR spectrum contains the following absorptions: 1330 (vs), 1310 (sh), 1170 (m), 1138 (vs), 1117 (vs), 1060 (m), 982 (m), 719 (m) and 704

---

*The checkers recommend the use of a "tipper tube" arrangement consisting of a 250-mL round-bottomed flask (containing the reactants) and an appropriate ground glass joint connecting adapter. The nitrogen flow is over the top of the reaction mixture (connected at the top of the condenser). They found that heating at reflux and stirring to dispel HCl was required for 10 hr, or for 3 hr followed by 24 hr at 45°. Filtration was accomplished with a sintered filter or the use of steel needle techniques. The Soxhlet extraction used 200 mL of THF and, after filtration to remove the first crop of product, the volume is slowly reduced to give more material. If the volume is reduced too much, however, the product may separate as an oil.

cm^{-1} (m). The mass spectrum shows the molecule ion at *m/e* 216 with an intensity of 4%.

Anal. For $S_4N_4O_2$: Calcd. N 25.9%, S 59.3%. Found N 25.7%, S 59.5%.

B. TRISULFURDINITROGEN OXIDE, S_3N_2O

$$S_3N_2Cl_2 + HCOOH \longrightarrow S_3N_2O + 2HCl + CO$$

■ **Caution.** *Avoid skin contact with formic acid. Serious skin damage might occur. Use rubber gloves. If contact has occurred remove contaminated clothes and rinse with plenty of water!*

A sample of $S_3N_2Cl_2$ (7.6 g 40 mmol) is placed in a 250 mL two-neck flask fitted with a reflux condenser topped with a $CaCl_2$ drying tube and with a glass inlet tube for a gentle nitrogen flow. Dry methylene chloride (150 mL) and 3 mL of formic acid are added under magnetic stirring. The mixture soon turns red and is gently refluxed until evolution of HCl stops. The solution is concentrated to $\frac{1}{4}$ its original volume and filtered under nitrogen through a glass-sintered funnel. The solvent is then evaporated *in vacuo.* Distillation of the liquid *in vacuo* affords 4.5 g (81%) of a red oil, bp 50° at $1{-}10^{-5}$ bar. The checkers provide the following details. Formic acid is dried by refluxing overnight and distillation from phthalic anhydride (equimolar proportions) followed by refluxing (4 hr) over and distillation from anhydrous copper(II) chloride. The reaction works best if CH_2Cl_2 is added first, followed by slow addition (over 2–3 hr) of formic acid. The refluxing was continued for 1 week before the evolution of HCl ceased. At this stage the reaction mixture has a dark red color. Distillation required an oil bath temperature of 60–70°. The yield was 57–70%.

■ **Caution.** *Do not distil to complete dryness and avoid bath temperatures above 100°! The solid residue might contain temperature-unstable byproducts that can explode violently.*

Properties

1-Oxo-1λ^4,2,4λ^4,3,5-trithiadiazole is a red, oily liquid, which does not wet glass. In the refrigerator, yellow needles are formed that melt at 18°. It can be stored under nitrogen without decomposition and is soluble in most common organic solvents. The IR spectrum shows absorptions at 1125 (vs), 980 (s), 903 (s), 734 (vs) and 663 cm^{-1} (s). The mass spectrum contains the molecule ion *m/e* 140 with an intensity of 11%.

Anal. For S_3N_2O: Calcd. N 20.0%, S 68.6%. Found: N 19.8%, S 69.0%.

C. DITHIADIAZOLONE, S_2N_2CO

$$S_4N_4 + [(CH_3)_3Sn]_3N \longrightarrow$$
$$2S_2N_2Sn(CH_3)_2 + Sn(CH_3)_4 + \tfrac{1}{3}N_2 + \tfrac{1}{3}(CH_3)_3N \quad (5)$$

$$S_2N_2Sn(CH_3)_2 + COF_2 \longrightarrow S_2N_2CO + (CH_3)_2SnF_2 \quad (6)$$

3.7 g (20 mmol) S_4N_4 is placed in a two-neck flask equipped with reflux cooler topped with a $CaCl_2$ drying tube and a dropping funnel (preferably with a Teflon valve) filled with 10.2 g (20 mmol) $[(CH_3)_3Sn]_3N$. The flask is cooled to $-15°$ and the amine slowly added. If solidification (mp 27°) occurs, the amine is gently warmed with a heat gun. After complete addition the resulting red liquid is heated to 90° until a yellow precipitate is formed (approximately 1 day). Volatiles then are removed *in vacuo* and the compound recrystallized from methylene chloride (200–400 mL). If the scale is increased, it is advisable to add some dry CCl_4 to the S_4N_4 and the amine, *because the reaction is extremely exothermic and the S_4N_4 might explode if the temperature rises above 100°*. The yield is 8.8 g (91%). (The checkers obtained a 73% yield. They report that the product is sometimes contaminated by S_8, which can be removed by washing twice with 30-mL portions of carbon disulfide.)

Properties

5,5-Dimethyl-1,3λ^4,2,4,5-dithiadiazastannole forms yellow crystals that are stable up to 198°, soluble only in methylene chloride, and stable toward water, but hydrolyzed by acids.[17] It is monomeric in the gas phase, and dimeric in solution and in the solid state. The single crystal X-ray structure determination shows the dimer having a Sn_2N_2 four-membered ring with C_i symmetry. The ring skeleton is nearly planar.[18] The ^1H NMR spectrum shows a singlet at $\delta = 0.95$ ppm with satellites $J_{H-^{117}Sn} = 66.3$, $J_{H-^{119}Sn} = 69.3$ Hz. The IR spectrum exhibits absorptions at 2960 (w), 2880 (w), 1393 (w), 1186 (w), 1064 (s), 1034 (vs), 901 (s), 778 (s), 737 (w), and 702 cm^{-1} (s). The molecule ion in the mass spectra is observed at m/e 242 (^{120}Sn) with an intensity of 60%. No higher peaks assignable to the dimer occur.

Anal. Calcd. for $C_2H_6N_2S_2Sn_2$: C 10.0%, H 2.5%, N 11.6%, S 26.5%, Sn 49.4%. Found: C 9.8%, H 2.4%, N 11.0%.

7.3 g (30 mmol) $S_2N_2Sn(CH_3)_2$ is suspended in 120-mL dry methylene chloride in a 250-mL two-neck flask with gas inlet tube, $CaCl_2$ drying tube, and a magnetic stirrer. (■ **Caution.** *Carbonyl fluoride is a toxic gas. Reactions should be carried out in a well-ventilated hood.*) The COF_2 cylinder is attached directly to the gas inlet tube and an excess of the gas is passed slowly through the solution. After 2 hr the solvent is removed under reduced pressure (>15 mbar). A sublimation finger is attached to the flask, the system cooled down with liquid air, evacuated to 10^{-5} bar, the stopcock closed and the flask warmed to 30–35°. The S_2N_2CO sublimes in yellow cubes in a yield of 1.8 g (50%).

Properties

$1,3\lambda^4,2,4$-Dithiadiazol-5-one forms yellow translucent crystals with the space group *Pbca-D*[15] and with all the bond lengths in the usual range.[11a] It melts at 40.5° and decomposes in contact with water. With moist air it forms a hydrate that can be dehydrated with P_4O_{10}. It is soluble in all organic solvents and in liquid SO_2. The IR spectrum reveals bands at 1727 (vs), 1172 (w), 1158 (w), 1065 (m), 780 (m), 725 (s), 640 (w), 520 (w), and 423 (w) cm^{-1}. The Raman spectrum shows absorptions at 1702 (w), 1500 (w), 1265 (w), 1065 (m), 905 (vs), 785 (s), 610 (w), 598 (s), 570 (w), 525 (s), and 425 (w) cm^{-1}. In the mass spectrum the molecule ion is found at m/e 120 with 12% intensity.

Anal. Calcd. for S_2N_2CO: C 10.0%, N 23.3%, S 53.4%. Found: C 10.0%, N 23.4%, S 52.9%.

References

1. A. Gregory, *J. Pharm.*, **21**, 315 (1935); **22**, 301 (1935).
2. E. Demarçay, *Compt. Rend.*, **91**, 854 (1880); **91**, 1066 (1880); **92**, 726 (1881).
3. H. W. Roesky, W. Grosse Böwing, I. Rayment, and H. M. M. Shearer, *J. Chem. Soc. Chem. Commun.* **1975**, 735.
4. H. W. Roesky and H. Wiezer, *Angew. Chem.*, **87**, 254 (1975); *Angew. Chem. Int. Ed. (Engl.)*, **14**, 258 (1975).
5. H. W. Roesky, W. Schaper, O. Petersen, and T. Müller, *Chem. Ber.*, **110**, 2695 (1977).
6. H. W. Roesky and E. Wehner, *Angew. Chem.*, **87**, 521 (1975); *Angew. Chem. Int. Ed. (Engl.)*, **14**, 498 (1975).
7. R. Neidlein and P. Leinberger, *Chem. Ztg.*, **99**, 433 (1975).
8. (a) H. W. Roesky, M. Witt, M. Diehl, J. W. Bats, and H. Fuess, *Chem. Ber.*, **112**, 1372 (1981). (b) H. W. Roesky, M. Witt, B. Krebs, and H. J. Korte, *Chem. Ber.*, **114**, 201 (1981). (c) H. W. Roesky, M. Witt, W. Clegg, W. Isenberg, M. Noltemeyer, and G. M. Sheldrick, *Angew. Chem.*, **92**, 959 (1980); *Angew. Chem. Int. Ed. (Engl.)*, **19**, 1943 (1980). (d) M. Witt and H.W. Roesky, *Z. Anorg. Allg. Chem.*, **51**, 515 (1984). (e) T. Chivers, A. W. Cordes, R. T. Oakley, and W. T. Pennington, *Inorg. Chem.*, **22**, 2429 (1983).
9. (a) H. W. Roesky, W. Clegg, J. Schimkowiak, M. Schmidt, M. Witt, and G. M. Sheldrick, *J. Chem. Soc. Dalton Trans.*, **1982**, 2117. (b) H. W. Roesky, M. Witt, J. Schimkowiak,

M. Schmidt, M. Noltemeyer, and G. M. Sheldrick, *Angew. Chem.*, **94**, 541 (1982); *Angew. Chem. Int. Ed. (Engl.)*, **21**, 536 (1982).
10. (a) H. W. Roesky, M. Kuhn, and J. W. Bats, *Chem. Ber.*, **115**, 3025 (1982). (b) H. W. Roesky, M. Thomas, J. Schimkowiak, M. Schmidt, M. Noltemeyer, and G. M. Sheldrick, *J. Chem. Soc. Chem. Commun.*, **1982**, 790. (c) H. W. Roesky, M. Thomas, J. W. Bats, and H. Fuess, *J. Chem. Soc. Dalton Trans.*, **1983**, 1891.
11. (a) H. W. Roesky, E. Wehner, E.-J. Zehnder, H.-J. Deiseroth, and A. Simon, *Chem. Ber.*, **111**, 1670 (1978). (b) A. Gieren, B. Dederer, R. Martin, F. Schanda, H. W. Roesky, and M. Eiser, *Chem. Ber.*, **113**, 3904 (1980). (c) H. W. Roesky, M. Thomas, M. Noltemeyer, and G. M. Sheldrick, *Angew. Chem.*, **94**, 861 (1982); *Angew. Chem. Int. Ed. (Engl.)*, **23**, 858 (1982).
12. W. L. Jolly and K. D. Maguire, *Inorg. Synth.*, **9**, 102, (1967).
13. (a) M. Becke-Goehring, *Inorg. Synth.*, **6**, 123, (1960). (b) G. Brauer, Handbuch d. präp. anorg. Chem. 403, F. Enke, Stuttgart 1975. (c) M. Villena-Blanco and W. L. Jolly, *Inorg. Synth.*, **9**, 98 (1967). (d) A. J. Banister, *Inorg. Synth,* **17**, 197 (1977).
14. (a) K. Sisido and S. Kozima, *J. Org. Chem.*, **29**, 907 (1964). (b) W. L. Lehn, *J. Am. Chem. Soc.*, **86**, 305 (1964).
15. M. Witt, Ph.D. Thesis, Frankfurt, Johann Wolfgang Goethe-Universität, Federal Republic of Germany, 1980.
16. P. G. Jones, W. Pinkert, and G. M. Sheldrick, *Acta Cryst.*, **C39**, 827 (1983).
17. H. W. Roesky and H. Wiezer, *Angew. Chem.*, **85**, 722 (1973); *Angew Chem. Int. Ed. (Engl.)*, **12**, 674 (1973).
18. H. W. Roesky, *Adv. Inorg. Chem. Radiochem.*, **22**, 239 (1979).

Chapter Two

INORGANIC POLYMER SYSTEMS

14. ORGANOSILANE HIGH POLYMERS: POLY(METHYLPHENYLSILYLENE)

$$\text{PhMeSiCl}_2 \xrightarrow[110°]{\text{Na, toluene}} (\text{PhMeSi})_n$$

Submitted by R. WEST* and P. TREFONAS†
Checked by W. P. WEBER‡ and Y.-X. DING‡

The first organosilane polymers based on diphenylsilylene units may have been synthesized by Kipping in 1924,[1] and polydimethylsilylene was first described by Burkhard in 1949.[2] These polymers were, however, highly crystalline, intractable white powders that decomposed when heated. Recently, with the synthesis of air-stable, soluble, formable organosilane polymers and copolymers, there has been a resurgence of interest in these materials.[3] The polysilanes have interesting electronic and conformational properties, and may be useful as precursors for β-SiC,[4] as impregnating agents for strengthening ceramics,[5] as photoresists for microelectronics,[6] as photoinitiators,[7] and as photoconductors.[8]

The solubility properties of polyorganosilanes are quite dependent on

*Department of Chemistry, University of Wisconsin, Madison, WI 53706.
†Monsanto Electronic Materials Co., 800 N. Lindbergh Blvd., St. Louis, MO 63167.
‡Department of Chemistry, University of Southern California, Los Angeles, CA 90089-1661.

the nature of the substituents bonded to the polymer backbone and on the molecular weight. Polymers with two or more different substituents, such as $(PhMeSi)_n$ or $(n\text{-}PrMeSi)_n$, or polymers with highly flexible substituents, such as $(n\text{-}HexMeSi)_n$ or $(n\text{-}Hex_2Si)_n$, are soluble in a wide variety of common organic solvents. Random copolymers, such as $[(cyclohexyl\text{-}MeSi)_x(n\text{-}HexMeSi)_y)]_n$ are also rather soluble. However, certain homopolymers with identical substituents, such as $(Me_2Si)_n$ or $(Ph_2Si)_n$, are crystalline and tend to be quite insoluble. Soluble copolymers containing units with identical substituents, for example, $[(PhMeSi)_x(Me_2Si)_y]_n$, display greatly enhanced solubility (although in this case some insoluble polymer containing long blocks of insoluble Me_2Si units is also formed). Decreasing the molecular weight of the polymers will also increase solubility; for example, $Me(Me_2Si)_{24}Me$ is somewhat soluble in several solvents.[9]

A. POLY(METHYLPHENYLSILYLENE)

$$PhMeSiCl_2 + 2Na \longrightarrow (PhMeSi)_n + 2NaCl$$

Procedure

The reaction is carried out in oven-dried glassware that has been thoroughly purged with dry nitrogen or argon. Pure dichlorosilane monomer is essential to produce a high molecular weight polymer. Because the polyorganosilanes are highly light sensitive, especially in solution, dark glassware or aluminum foil wrap should be used to protect the polymer from light during the reaction and subsequent work-up.

Purify commercially obtained $PhMeSiCl_2$* by careful fractional distillation through a 15-cm jacketed fractionating column filled with glass helices. Equip a 2-L, three-necked, round-bottom flask with a 250-mL pressure-equalizing dropping funnel, a high-capacity reflux condenser topped with a nitrogen by-pass, a high-speed overhead mechanical stirrer, and a heating mantle. To the flask add 1 L of toluene, dried by refluxing over sodium, and 47.1 g (2.05 mol) of fresh lump Na cut into small pieces. Cannulate (or syringe) into the dropping funnel 161 mL (1.00 mol) of distilled $PhMeSiCl_2$. Bring the toluene to reflux and stir the reaction at high speed to produce finely divided molten Na. Then add the $PhMeSiCl_2$ slowly over a 1.5-hr period, maintaining the toluene at gentle reflux throughout the addition by adjusting the heating mantle if necessary.

---

*Available from Petrarch Systems, Inc., Bartram Road, Bristol PA 19007.

(■ **Caution.** *This reaction is highly exothermic. Do not add the dichlorosilane rapidly!*) The reaction mixture should turn a deep purple color. After the addition is complete, reflux for an additional 1.5 hr and then let the reaction cool to room temperature.

Prepare a slurry of 5 g of NaHCO$_3$ in 50 mL of 2-propanol, and add it *very slowly*, with stirring, to the reaction mixture to destroy the excess Na. (The mixture may foam during the addition; be prepared to collect any foam that rises out of the condenser.) Stir for an additional 2 hr to insure that all the Na was consumed.

Reduce the volume of the reaction mixture to about 500 mL by rotary evaporation. Add the mixture slowly to a large flask containing well-stirred 2-propanol, 12 to 15 times the volume of the reaction mixture. The polymer and most of the NaCl will precipitate, leaving the oligomers dissolved in the 2-propanol–toluene solution. Collect the polymer–NaCl precipitate by filtration and set it aside until it is completely dry. Then redissolve the polymer by stirring for 24 hr in 1 L of toluene. (If the viscosity is too high, more toluene can be added.) Extract this solution with several 1 L volumes of water to remove the salts. After the purple color is discharged, extract twice more with water. If an emulsion forms, it can be broken up with a *wooden stick*, which will penetrate the bubbles. Distill off the toluene using a rotary evaporator and dry the polymer in a vacuum oven at 50° for 48 hr. Yield: 66.5 g (55%).

Anal. Calc. for C$_7$H$_8$Si: C, 69.9; H, 6.7; Si, 23.4. Found: C, 69.4; H, 6.7; Si, 23.1.

The oligomers can be isolated by removing the solvent from the 2-propanol–toluene solution by rotary evaporation, redissolving in hexane and extracting the hexane solution with water to remove any residual salts. Then dry the solution over MgSO$_4$ and isolate the pure oligomers by removing the hexane solvent by rotary evaporation. Yield: 48.7 g (41%).

Properties

The poly(methylphenylsilylene) is obtained as a clear, colorless, brittle air-stable solid that is soluble in benzene, toluene, THF, and chloroform. (*Note:* Chloroform solutions are especially light sensitive.) Molecular weight distributions can be determined by gel permeation chromatography using THF or toluene as the eluent. Typically, the molecular weight distribution is bimodal, with $\overline{M}_w$ of ~200,000 (relative to polystyrene standards) for the high molecular weight fraction of the bimodal distribution, and $\overline{M}_w$ of ~6000 for the low molecular weight fraction of the distribution.

The IR spectrum contains bands at 1249 cm^{-1} (Si-methyl), 1101 cm^{-1}

(Si-phenyl), 2960 and 2900 cm^{-1} (C—H, aliphatic), 3051 and 3065 cm^{-1} (C—H aromatic). The ^{1}H NMR spectrum has resonances at $\delta = 0.17$ (broad, CH$_3$) and $\delta = 7.38$ (broad, C$_6$H$_5$). The ^{13}C NMR spectrum has broad resonances at $\delta = -6.7$ to -5.4, 127.6 to 129.3, and 135.0 to 136.3. The UV spectrum for high molecular weight poly(methylphenylsilylene) in THF has an absorption maximum at 341 nm.*

B. PREPARATION OF OTHER POLYORGANOSILANES

Other polyorganosilanes can be prepared by a similar procedure. For polymers with *n*-alkyl substituents, hexane can be substituted for toluene in the work-up. For (*cyclo*-C$_6$H$_{11}$SiCH$_3$)$_n$, which is soluble in only hot toluene or cyclohexane, the work-up is simplified by substituting cyclohexane for toluene when redissolving the polymer. Copolymers can be prepared by the same procedure as homopolymers, by using a mixture of the two monomers in the dropping funnel at the start of the reaction. The yields of alkyl-substituted polyorganosilanes tend to be lower than those for aryl-substituted polymers, often only 10 to 35%, depending on the monomer used. Copolymers with alkyl substituents are usually formed in higher yields, from 30 to 70%.

The use of 25% 1,1'-oxybis(2-methoxyethane) (diglyme)/75% toluene as the solvent mixture for the polymerization has been reported to give increased yields of alkyl-substituted polyorganosilanes, although with lower molecular weight.[8]

References

1. F. S. Kipping, *J. Chem. Soc.*, **125**, 2291 (1924).
2. C. A. Burkhard, *J. Am. Chem. Soc.*, **71**, 963 (1949).
3. For a review on polysilanes see R. West, *J. Organomet. Chem.*, **300**, 327–346 (1986).
4. R. West, L. D. David, P. I. Djurovich, Chapter 19 and H. Yu, *J. Am. Cer. Soc.*, **62**, 899 (1983); R. West, *Ultrastructure Processing of Ceramics, Glasses and Composites*, L. L. Hench and D. R. Ulrich (eds.), Wiley, New York, 1984, Chapter 19; R. H. Baney, Chap. 20, *ibid.*, C. H. Beatty, *ibid.*, Chap. 22.
5. K. S. Mazdyansi, R. West, and L. D. David, *J. Am. Cer. Soc.*, **61**, 504 (1978).
6. R. D. Miller, D. Hofer, D. R. McKean, C. G. Willson, R. West, and P. Trefonas III, in *Materials for Microlithography*, C. G. Willson and J. M. J. Fréchet (eds.) (ACS Symposium Series), American Chemical Society, Washington, DC, 1984, Chap. 14; J. M. Zeigler, L. A. Harrah, and A. W. Johnson, *SPIE, Advances in Resist Technology*, **539**, 166 (1985).

*The checkers report λ_{max} 338 nm, and a higher yield of polymer (70%), probably because a small amount of oligomer was precipitated with the polymer.

7. R. West, A. R. Wolff, and D. G. Peterson, *J. Radiat. Curing.* **13,** 35 (1986); A. R. Wolff and R. West, *Appl. Organomet. Chem.,* **1,** 7 (1987).
8. R. G. Kepler, J. M. Zeigler, L. A. Harrah, and S. R. Kurtz, *Phys. Rev. B.* **35,** 2818 (1987); M. Stolka, H.-J. Yah, K. McGrane, and D. M. Pai, *J. Polym. Sci., Polym. Chem. Ed.,* **25,** 823 (1987); M. Fujino, *Chem. Phys. Lett.,* **136,** 451 (1987).
9. W. Boberski and A. L. Allred, *J. Organomet. Chem.,* **88,** 65 (1975).

15. ORGANOSILICON DERIVATIVES OF CYCLIC AND HIGH POLYMERIC PHOSPHAZENES

Submitted by DAVID J. BRENNAN,* JAMES M. GRAASKAMP,*
BEVERLY S. DUNN,* and HARRY R. ALLCOCK*
Checked by MICHAEL SENNETT†

Species that combine the properties of organosilicon compounds and phosphazenes are prepared by the linkage of organosilicon side groups to a small molecule cyclic or linear high polymeric phosphazene skeleton. This is particularly important for high polymeric derivatives in which hybrid properties typical of polysiloxanes (silicones) and polyphosphazenes[1,2] may be obtained.

The synthetic route to such high polymers involves first the linkage of an organosilicon unit to a small molecule cyclic phosphazene (in this case, a cyclo*tri*phosphazene). The linkage is accomplished by the use of an organosilyl Grignard reagent to replace chlorine atoms in a chlorocyclophosphazene. Further structural diversity is accomplished by replacement of a chlorine atom geminal to the organosilicon group by a methyl group, again with the use of the appropriate Grignard reagent. For mechanistic reasons not yet fully understood, the presence of the methyl groups aids the next step, which is the thermal ring-opening polymerization of the cyclotriphosphazene to a high molecular weight linear polymer. The remaining chlorine atoms are then replaced by trifluoroethoxy units to eliminate the hydrolytic sensitivity that accompanies the presence of P—Cl bonds. This chlorine replacement step leaves the organosilicon groups intact if it is carried out in toluene as a solvent. However, if THF is employed as the solvent, attack by trifluoroethoxide ion on a —CH_2-$SiMe_3$ group brings about cleavage of a carbon–silicon bond leaving a methyl group attached to phosphorus.

Although a variety of organosilicon groups have been linked to phos-

*Department of Chemistry, The Pennsylvania State University, University Park, PA 16802.
†U.S. Army Materials Technology Laboratory, Watertown, MA 02172-0001.

phazenes,[3-6] use of the trimethylsilylmethyl unit has been studied in the greatest detail and is described here.

All reactions and manipulations are carried out under a nitrogen atmosphere up to the point at which the trifluoroethoxy groups have been attached to the phosphazene skeleton. The glassware used during Grignard and trifluoroethoxide reactions is dried in an oven at 130° for 24 hr and is cooled under vacuum or nitrogen before use.

A. *gem*(METHYL-TRIMETHYLSILYLMETHYL) TETRACHLOROCYCLOTRIPHOSPHAZENE*

Procedure

$$Me_3SiCH_2Cl + Mg \xrightarrow{THF} Me_3SiCH_2MgCl$$

■ **Caution.** *An ice bath should be ready at hand to control the exothermic Grignard reaction if necessary.*

Magnesium (3.65 g, 0.150 mol) and 200 mL of dry THF (distilled over sodium benzophenone) are added to a 500-mL three-necked round-bottom flask equipped with a magnetic stirring bar, a rubber septum, and a reflux condenser fitted with a nitrogen inlet adapter. To the unstirred reaction mixture 0.5–1.0 mL of 1,2-dibromoethane is added directly to the magnesium via syringe. Within 15–20 min, bubbles of ethylene are evolved to indicate that the magnesium is activated. In a nitrogen filled glove bag, 13.8 mL (0.100 mol) of dry Me_3SiCH_2Cl (dried over activated 3-Å molecular sieves) is drawn into a dry syringe fitted with a Luer lock stopcock

*2,2,4,4-Tetrachloro-6-methyl-6-[(trimethylsilyl)methyl]-1,3,5,2λ^5,4λ^5,6λ^5-triazatriphosphorine.

and syringe needle. The Me_3SiCH_2Cl is then added to the magnesium at the bottom of the flask. The mixture is allowed to stand for 10 min and is then stirred. A vigorous reaction will occur within 30 min, causing the solvent to reflux. The reaction mixture is allowed to reflux and should be cooled only if the solvent begins to reflux past the water-cooled condenser. The reaction mixture is then stirred until it cools to room temperature.

A 1-L three-necked, round-bottom flask is equipped with a reflux condenser fitted with a nitrogen inlet adapter, a 250-mL pressure equalizing addition funnel, and a magnetic stirbar. To this reaction vessel is added 30.0 g (0.0865 mol) of $(NPCl_2)_3$ and 500 mL of dry THF. The THF solution of Me_3SiCH_2MgCl is transferred to the addition funnel via a double-tipped syringe needle. The solution of $(NPCl_2)$ in THF is heated to reflux (66°), and the Grignard reagent is added dropwise over a 1–2 hr period. On completion of addition, the reaction mixture is stirred at 66° for 24 hr, then cooled to room temperature.

In a nitrogen-filled glove bag, 45 mL of 2.9 M CH_3MgCl (0.13 mol) in THF is transferred to an addition funnel, which is then attached to a 1-L flask. The CH_3MgCl solution is then added dropwise to the THF solution of $N_3P_3Cl_5(CH_2SiMe_3)$, which has been cooled to 0° with an ice–water bath. On completion of addition, the solution is allowed to warm to 25° and is stirred for 16 hr.

The solvent is removed from the reaction mixture under reduced pressure on a rotary evaporator in a 1-L recovery flask to yield a white residue. (■ **Caution.** *Washing of the mixture with 5% HCl in water results in the solution of magnesium salts: This is an exothermic process. Care should be taken to prevent the diethyl ether from boiling. Use chilled 5% HCl in water solution.*) To the residue is added 500 mL of diethyl ether and 250 mL of 5% HCl, which has been chilled at 0°. Gentle swirling yields a two-phase mixture, which is transferred to a 1-L separatory funnel. The aqueous (lower) phase is separated and the ethereal layer is washed with two 250-mL portions of 5% HCl. The ethereal layer is dried over $MgSO_4$ and suction filtered through Fuller's Earth to yield a clear filtrate. The diethyl ether is removed under reduced pressure to yield crude *gem*-$N_3P_3Cl_4(CH_3)(CH_2SiMe_3)$ as a solid. The white solid is recrystallized twice from hot hexane, and is vacuum sublimed at 100° (0.05 torr) twice to yield pure *gem*-$N_3P_3Cl_4(CH_3)(CH_2SiMe_3)$, mp 114–115°. Yield: 19.7 g, 60%.*

Anal. Calcd for *gem*-$N_3P_3Cl_4(CH_3)(CH_2SiMe_3)$: C, 15.84; H, 3.73; N, 11.09. Found: C, 15.87; H, 3.62; N, 10.98.

---

*The checker reports that considerable loss of material occurred during the purification process and that only a 35% yield of compound melting at 114° was obtained. Use of spectroscopic grade *n*-hexane as a recrystallization solvent improved the yield.

Properties

The trimer, gem-$N_3P_3Cl_4(CH_3)(CH_2SiMe_3)$, is a white solid that is soluble in THF, diethyl ether, 1,4-dioxane, CH_2Cl_2, $CHCl_3$, benzene, acetone, ethanol, toluene, and hot hexane. It is insoluble in water. The compound is relatively stable in the solid state in the presence of atmospheric moisture, but hydrolyzes slowly over a period of months or years. For this reason, the compound is stored under an atmosphere of dry nitrogen or in sealed vials under vacuum. The compound readily sublimes at 100° (0.05 torr), which makes it readily purified for polymerization purposes. The ^{31}P NMR spectrum ($CDCl_3$) is an AB_2 spin pattern (ν_A: 38.0 ppm, ν_B: 16.9 ppm, $J_{PNP} = 5.8$ Hz). The ^1H NMR spectrum ($CDCl_3$) shows resonances for PCH_3 at $\delta = 1.69$ (d) 3H ($J_{PCH} = 14.1$ Hz), $PCH_2Si(CH_3)_3$ at $\delta = 1.31$ (dt) 2H ($J_{PCH} = 17.2$ Hz, $J_{PNPCH} = 3.9$ Hz), and $PCH_2Si(CH_3)_3$ at $\delta = 0.23$ (s) 9H. The ^{13}C NMR spectrum ($CDCl_3$) shows resonances for PCH_3 at $\nu = 23.33$ ppm (dt) ($J_{PC} = 88.9$ Hz, $J_{PNPC} = 2.5$ Hz), $PCH_2Si(CH_3)_3$ at $\nu = 23.16$ ppm (dt) ($J_{PC} = 88.6$ Hz, $J_{PNPC} = 5.1$ Hz), and $PCH_2Si(CH_3)_3$ at $\nu = -0.15$ ppm (d) ($J_{PCSiC} = 3.4$ Hz). The mass spectrum showed a parent ion peak at 379 amu with a Cl_4 isotope pattern. Major fragmentation patterns were loss of methyl (M − 15) and chlorine (M − 35).

B. POLY[gem(METHYLTRIMETHYLSILYLMETHYL)-CHLOROPHOSPHAZENE]*

The trimer, gem-$N_3P_3Cl_4(CH_3)(CH_2SiMe_3)$, is purified rigorously by three additional vacuum sublimations at 100° (0.05 torr) and is stored in a nitrogen atmosphere before use. Into a clean, dry, Pyrex glass tube (with a length of 200 mm, o.d. of 20 mm, wall thickness of 1 mm, with a constriction 40 mm long and 30 mm from the open end) is added 10.0 g (0.0270 mol) of gem-$N_3P_3Cl_4(CH_3)(CH_2SiMe_3)$ under an atmosphere of nitrogen. The tube is evacuated for a period of 30 min at 0.05 torr, then sealed at the constriction. This glass ampule is wrapped in a wire mesh screen and heated at 210° in a thermoregulated oven (with a mechanical tube rocker) until

*Poly[2,2,4,4-tetrachloro-6-methyl-6[(trimethylsilyl)methyl]catenatriphosphazene-1,6-diyl].

the contents of the tube reach maximum viscosity but are not immobile (2–6 hr).*

■ **Caution.** *The tube is wrapped in paper towels to prevent ripping of the glove bag when being opened.*

The tube is removed from the oven, allowed to cool, then opened (shattered by percussion) in a nitrogen-filled glove bag. The contents of the tube and shattered glass are transferred to a 500-mL airless flask equipped with a magnetic stir bar. The conversion of $gem\text{-}N_3P_3Cl_4(CH_3)(CH_2SiMe_3)$ to the high polymer is approximately 50–60%.

Properties

The polymer $\{[NPCl_2]_2\text{—}[NP(CH_3)(CH_2SiMe_3)]\}_n$ is readily soluble in dry THF, 1,4-dioxane, benzene, and toluene. It crosslinks readily to an insoluble gel in contact with moisture due to the presence of hydrolytically unstable P—Cl bonds. Extreme care must be taken to avoid contact with moisture during the manipulations of the polymer. Residual cyclic trimer can be separated from the macromolecular species after treatment of the polymer–trimer mixture with sodium trifluoroethoxide as described in the next two sections. The ^{31}P NMR spectrum of $\{[NPCl_2]_2\text{—}[NP(CH_3)(CH_2SiMe_3)]\}_n$ (THF, D_2O lock) from the polymer–trimer mixture shows two resonances at 18.9 and -28.6 ppm for $P(CH_3)(CH_2SiMe_3)$ and PCl_2, respectively. The ^1H NMR spectrum of the polymer–trimer mixture (C_6D_6) shows distinct resonances for protons of the trimer and polymer. Resonances for protons in the trimer are detected for PCH_3 at $\delta = 1.20$ (d) 3 H$(J_{PCH} = 14.1$ Hz$)$ $PCH_2Si(CH_3)_3$ at $\delta = 0.70$ (dt) 2H $(J_{PCH} = 17.5$ Hz, $J_{PNPCH} = 3.6$ Hz$)$, and $PCH_2Si(CH_3)_3$ at $\delta = $ (s) 9H. Resonances for the protons of the high polymer are detected for PCH_3 at $\delta = 1.60$ (d) 3H $(J_{PCH} = 13.8$ Hz$)$, $PCH_2Si(CH_3)_3$ at $\delta = 1.20$ (d) 2H $(J_{PCH} = 14.1$ Hz$)$, and $PCH_2Si(CH_3)_3$ at $\delta = 0.21$ (s) 9H.

C. POLY[gem(METHYLTRIMETHYLSILYLMETHYL)-TRIFLUOROETHOXYPHOSPHAZENE]†

Procedure

$$CF_3CH_2OH + Na \xrightarrow{\text{toluene}} CF_3CH_2ONa + \tfrac{1}{2}H_2$$

---

*The checker comments that this polymerization was somewhat unpredictable and that the viscosity increase was negligible during the first 3–4 hr, but then accelerated rapidly during the next 30 min. Thus, constant monitoring was needed.

†Poly[2-methyl-4,4,6,6-tetrakis(2,2,2-trifluoroethoxy)-2-[(trimethylsilyl)methyl]-catenatriphosphazene-1,6-diyl].

$$\left[\left(-N{=}\underset{\underset{\displaystyle Cl}{|}}{\overset{\overset{\displaystyle Cl}{|}}{P}}-\right)_{2}\left(-N{=}\underset{\underset{\displaystyle CH_3}{|}}{\overset{\overset{\displaystyle CH_2SiMe_2}{|}}{P}}-\right)\right]_n + 4CF_3CH_2ONa \xrightarrow[110°]{\text{toluene}}$$

$$\left[\left(-N{=}\underset{\underset{\displaystyle OCH_2CF_3}{|}}{\overset{\overset{\displaystyle OCH_2CF_3}{|}}{P}}-\right)_{2}\left(-N{=}\underset{\underset{\displaystyle CH_3}{|}}{\overset{\overset{\displaystyle CH_2SiMe_3}{|}}{P}}-\right)\right]_n + 4NaCl$$

To the polymer $\{[NPCl_2]_2{-}[NP(CH_3)(CH_2SiMe_3)]\}_n$ and the residual cyclic trimer, *gem*-$N_3P_3Cl_4(CH_3)(CH_2SiMe_3)$, in an airless flask is added 300 mL of dry toluene (distilled over sodium benzophenone). This mixture is stirred under an atmosphere of dry nitrogen until the contents of the broken tube dissolve (3–4 hr). The toluene solution is then transferred to a 500-mL addition funnel under a nitrogen atmosphere via a double-tipped syringe needle.

■ **Caution.** *Under no circumstances should mixtures that contain un-quenched CF_3CH_2ONa be evaporated to dryness and heated, otherwise a potentially violent, exothermic solid state reaction may occur. Addition of Me_3SiCl before removal of solvents brings about deactivation of CF_3CH_2ONa.*

To a 2-L, three-necked, round-bottom flask equipped with a condenser and fitted with a nitrogen inlet adapter and a magnetic stir bar is added 7.3 g (0.32 mol) of sodium and 1 L of dry toluene. In a nitrogen-filled glove bag, 26 mL (0.36 mol) of dry CF_3CH_2OH (dried and stored over activated 3-Å molecular sieves) is transferred to a pressure equalizing ad-dition funnel. The addition funnel is attached to the 2-L flask and the CF_3CH_2OH is added dropwise over a 30-min period. On completion of addition, the reaction mixture is stirred at 25° for 1 hr, then refluxed at 110° until foaming ceases (4–5 hr) to ensure that complete formation of CF_3CH_2ONa has occurred (CF_3CH_2ONa is insoluble in toluene at 25° but is soluble in hot toluene). The CF_3CH_2ONa solution is allowed to cool and the addition funnel containing the polymer–trimer mixture is attached to the 2-L flask. The CF_3CH_2ONa–toluene slurry is heated to reflux, which brings about solution of the CF_3CH_2ONa. The polymer–trimer solution is then added dropwise to the hot CF_3CH_2ONa solution. On completion of addition, the reaction mixture is heated at 110° for 10 hr. During this period of time, $\{[NP(OCH_2CF_3)_2]_2{-}[NP(CH_3)(CH_2SiMe_3)]\}_n$ precipitates from the hot toluene solution.

The reaction mixture is allowed to cool to 25° and the insoluble materials are allowed to settle. The supernatant toluene solution is removed via double-tipped syringe needle, and a solution of 40 mL (0.24 mol) of Me_3SiCl in 1 L of dry THF is added to the residue (Me_3SiCl reacts with residual CF_3CH_2ONa to form $Me_3SiOCH_2CF_3$ and NaCl). This mixture is stirred under nitrogen until the polymeric material dissolves (4–5 hr). The THF solution is transferred to a 1-L recovery flask in two portions, and the solvent is removed under reduced pressure by the use of a rotary evaporator in a hood. To the dry polymer–salt residue is added 500 mL of acetone and the mixture is shaken or stirred until the polymeric material dissolves (6–12 hr) to leave a milky solution. The acetone solution is concentrated to approximately 75 mL and the polymer is precipitated from acetone into water. The polymeric material is recovered and dried *in vacuo*. The acetone–water reprecipitation procedure is repeated twice more to remove the residual salts. The polymer is then reprecipitated from a concentrated acetone solution into pentane (twice), followed by Soxhlet extraction with pentane for 48 hr to remove residual cyclic trimers. The polymer is dried *in vacuo* for 24 hr to yield 4.3 g (25%) (13.6% yield by the checker) of a light brown elastomeric material.

Anal. Calcd: C, 24.64; H, 3.48; N, 6.64, Cl, 0.00. Found: C, 24.20; H, 3.21; N, 6.63, Cl, 0.17.

Properties

The polymer is a light brown, film-forming elastomer, which is soluble in THF, acetone, and methyl ethyl ketone, and is insoluble in H_2O, diethyl ether, 1,4-dioxane, CH_2Cl_2, toluene, benzene, hexane, and ethanol. Its glass transition temperature is $-55°$, which means that it is an elastomer down to this temperature. The weight average (M_w) and number average (M_n) molecular weights were estimated by gel permeation chromatography (GPC) in THF [with 0.1% (n-Bu)$_4$NBr] using polystyrene standards. The value for M_w is 7.5×10^5 and M_n is 8.6×10^4. The polymer has a very broad molecular weight distribution with a polydispersity value of 8 to 9. The ^{31}P NMR spectrum (THF, D_2O lock) shows resonances for $P(CH_3)(CH_2SiMe_3)$ at $+18.5$ ppm (m) and $P(OCH_2CF_3)_2$ at -9.2 ppm (m). The ^1H NMR spectrum (acetone-d_6) shows resonances for PCH$_3$ at $\delta = 1.71$ (d) ($J_{PCH} = 14.0$ Hz), PCH$_2$Si(CH$_3$)$_3$ at $\delta = 1.43$ (d) 2H ($J_{PCH} = 18.0$ Hz), PCH$_2$Si(CH$_3$)$_3$ at $\delta = 0.19$ (s) 9H, and $P(OCH_2CF_3)_2$ at $\delta = 4.48$ (m) 8 H.

D. POLY[*gem*(DIMETHYL)TRIFLUOROETHOXYPHOS-PHAZENE]*

Procedure

$$CF_3CH_2OH + Na \xrightarrow{THF} CF_3CH_2ONa + \tfrac{1}{2} H_2$$

$$\left[\left(\!\!\begin{array}{c} Cl \\ | \\ N{=}P \\ | \\ Cl \end{array}\!\!\right)_{\!2}\!\!\left(\!\!\begin{array}{c} CH_2SiMe_3 \\ | \\ N{=}P\text{---} \\ | \\ CH_3 \end{array}\!\!\right)\right]_n + 5CF_3CH_2ONa \xrightarrow[CF_3CH_2OH]{THF}$$

$$\left[\left(\!\!\begin{array}{c} OCH_2CF_3 \\ | \\ N{=}P\text{---} \\ | \\ OCH_2CF_3 \end{array}\!\!\right)_{\!2}\!\!\left(\!\!\begin{array}{c} CH_3 \\ | \\ N{=}P\text{---} \\ | \\ CH_3 \end{array}\!\!\right)\right]_n + CF_3CH_2OSiMe_3 + 4NaCl$$

To the polymer $\{[NPCl_2]_2\text{---}[NP(CH_3)(CH_2SiMe_3)]\}_n$ and the residual cyclic species, *gem*-$N_3P_3Cl_4(CH_3)(CH_2SiMe_3)$, in an airless flask is added 300 mL of dry THF (distilled over sodium benzophenone ketyl). This mixture is stirred under an atmosphere of dry nitrogen until the contents of the broken tube dissolve (3–4 hr). The THF solution is then transferred to a 500-mL addition funnel under an atmosphere of nitrogen via a double-tipped syringe needle.

To a 1-L, three-necked, round-bottom flask equipped with a reflux condenser fitted with a nitrogen inlet adapter and a magnetic stirring bar is added 9.4 g (0.41 mol) of Na and 500 mL of dry THF. In a nitrogen filled glove bag 33 mL (0.45 mol) of dry CF_3CH_2OH is transferred to a pressure equalizing addition funnel. (■ **Caution.** *The reaction between CF_3CH_2OH and sodium in THF is exothermic. A slow dropwise addition of CF_3CH_2OH to the THF–sodium mixture at $0°$ will prevent a violent reaction.*) The addition funnel is attached to the 1-L flask, and the CF_3CH_2OH is added dropwise over a 1-hr period to the sodium–THF mixture that has been cooled to $0°$. The reaction mixture is stirred until all of the sodium has reacted and no further evolution of H_2 can be detected.

The addition funnel containing the polymer–trimer mixture in THF is attached to the 1-L flask, and the CF_3CH_2ONa solution is heated to reflux

*Poly[2,2-dimethyl-4,4,6,6-tetrakis(2,2,2-trifluoroethoxy)catenatriphosphazene-1,6-diyl].

(66°). The polymer–trimer solution is added dropwise to the hot THF solution over a 2-hr period. On completion of addition the reaction mixture is heated at 66° for 24 hr.

■ **Caution.**   *Under no circumstances should mixtures that contain un-quenched CF_3CH_2ONa be evaporated to dryness and heated, otherwise a potentially violent, exothermic solid state reaction may occur. Addition of Me_3SiCl before removal of solvents brings about deactivation of CF_3CH_2ONa.*

The reaction mixture is allowed to cool to 25° and Me_3SiCl is added slowly ($\sim$50 mL) to the THF solution until the solution is just acidic when tested with wet litmus. The solution is then transferred to a 1-L recovery flask. The solvent and volatile species are removed under reduced pressure by use of a rotary evaporator in a hood. To the dry polymer–salt residue is added 500 mL of acetone and the mixture is shaken or stirred until the polymeric material dissolves (6–12 hr) to yield a milky solution. From this point on, the polymer is purified and isolated as described in the previous section. Yield: 3.8 g (25%) (79% yield by the checker).

Anal. Calcd: C, 21.40; H, 2.52; N, 7.49; Cl, 0.00. Found: C, 21.51; H, 2.65; N, 7.62; Cl, 0.10.

Properties

The polymer is a light brown, film-forming elastomer that is soluble in THF, acetone, and methyl ethyl ketone, and insoluble in H_2O, diethyl ether, 1,4-dioxane, CH_2Cl_2, toluene, benzene, hexane, and ethanol. Its glass transition temperaure is $-58°$. The M_w is 8.8×10^5 and M_n is 5.3×10^4. The polymer has a very broad molecular weight distribution with a polydispersity value between 16 and 17. The ^{31}P NMR spectrum (THF, D_2O lock) shows resonances for $P(CH_3)_2$ at $+15.6$ ppm(m) and $P(OCH_2CF_3)_2$ at -7.6 ppm(m). The ^1H NMR spectrum (acetone d_6) shows resonances for $P(CH_3)_2$ at $\delta = 1.67$ (d) 6 H ($J_{PCH} = 14.3$ Hz) and $P(OCH_2CF_3)_2$ at $\delta = 4.48$ (m) 8 H.

References

1. H. R. Allcock, *Phosphorus–Nitrogen Compounds*, Academic Press, New York, 1972.
2. H. R. Allcock, R. L. Kugel, and K. J. Valan, *Inorg. Chem.*, **5**, 1716 (1966).
3. H. R. Allcock, D. J. Brennan, J. M. Graaskamp, and M. Parvez, *Organometallics* **5**, 2434 (1986).
4. H. R. Allcock, D. J. Brennan, B. S. Dunn, and M. Parvez, *Inorg. Chem.* (1988).
5. H. R. Allcock, D. J. Brennan, and J. M. Graaskamp, *Macromolecules* **21**, 1 (1988).
6. H. R. Allcock, D. J. Brennan, and B. S. Dunn, *Macromolecules* (in press).

16. POLY(DIMETHYLPHOSPHAZENE) AND POLY(METHYLPHENYLPHOSPHAZENE)
{Poly[nitrilo(dimethylphosphoranylidyne)] and Poly[nitrilo(methylphenylphosphoranylidyne)]}

Submitted by P. WISIAN-NEILSON* and R. H. NEILSON†
Checked by JAMES M. GRAASKAMP‡ and BEVERLY S. DUNN‡

Amino, alkoxy, and aryloxy polyphosphazenes are typically prepared by nucleophilic displacement reactions of poly(dihalophosphazenes).[1] Analogous reactions with organometallic reagents, however, result in chain degradation and cross linking rather than in linear, alkyl, or aryl substituted poly(phosphazenes).[2] The thermolysis of appropriate silicon–nitrogen–phosphorus compounds[3,4] can be used to prepare fully P—C bonded poly(organophosphazenes).[5,6] The synthesis of two of these materials and their Si—N—P precursors is described here.

■ **Caution.** *The phosphines and several other reagents have obnoxious odors and are either known to be or are likely to be toxic. All reactions and manipulations should be carried out in efficient hoods.*

A. POLY(DIMETHYLPHOSPHAZENE)

Procedure

1. *P,P*-Dimethyl-*N,N*-bis(trimethylsilyl)phosphinous Amide

$$(Me_3Si)_2NH \xrightarrow[\text{(2) PCl}_3]{\text{(1) } n\text{-BuLi}} (Me_3Si)_2NPCl_2$$

$$(Me_3Si)_2NPCl_2 \xrightarrow{\text{2MeMgBr}} (Me_3Si)_2NPMe_2$$

As described in the Wilburn procedure,[3] $(Me_3Si)_2NH$ (104.3 mL, 0.50 mol) and Et_2O (500 mL, distilled from CaH_2) are placed in a 2-L, three-necked, round-bottom flask that is equipped with a nitrogen inlet, mechanical stirrer, and a pressure equalizing addition funnel fitted with a rubber septum. Then *n*-butyllithium (*n*-BuLi) (313 mL, 0.50 mol, 1.6 *M* in hexane) is transferred to the addition funnel via a cannula and added slowly (~0.5

*Department of Chemistry, Southern Methodist University, Dallas, TX 75275.
†Department of Chemistry, Texas Christian University, Forth Worth, TX 76129.
‡Department of Chemistry, The Pennsylvania State University, University Park, Pennsylvania 16802.

hr) to the stirred $(Me_3Si)_2NH$ solution that is cooled in an ice–water bath. When addition is complete, the mixture is stirred at room temperature for 1 hr and then cooled to $-78°$. Dry Et_2O (30–50 mL) is added to the mixture via the addition funnel in order to remove residual *n*-BuLi from the funnel. Keeping the solution at $-78°$, PCl_3 (43.6 mL, 0.50 mol) is added over 15 min, the mixture is stirred for 10 min at $-78°$, the dry ice–acetone bath is removed, and stirring is continued for 1 hr. This mixture is cooled to $0°$ and a cannula is used to transfer $MeMgBr$ (333 mL, 1.0 mol, 3.0 M in Et_2O) to the addition funnel after it has been rinsed as just described. The Grignard reagent is added over 1.5 to 2 hr and the mixture is then stirred at room temperature for 2 or 3 hr and allowed to stand overnight. Hexane (~300 mL) is added to facilitate precipitation of the salts. After filtration (Celite® filter aid is recommended by the checkers for all filtrations) under nitrogen and washing the residue with hexane (3 × 150 mL), the combined filtrate and washings are concentrated by removal of solvents at reduced pressure. If solids remain in the residue, more hexane should be added and the mixture filtered again. Solvent removal from the filtrate and distillation of the residue at 3.0 torr (bp 55°) affords $(Me_3Si)_2NPMe_2$ as a pungent, moisture-sensitive liquid. Yield: 75.6 g, 68%. ^{31}P NMR $\delta = 31.7$ $(CDCl_3)$.

■ **Caution.** *The salts from this reaction are best destroyed by slow, careful addition of isopropyl alcohol, followed by H_2O, and then dilute bleach to destroy the residual phosphinous amide.*

2. *P,P*-Dimethyl-*N*-(trimethylsilyl)phosphorimide Bromide

$$(Me_3Si)_2NPMe_2 + Br_2 \longrightarrow Me_3SiN{=}\overset{\overset{\displaystyle Br}{|}}{P}Me_2 + Me_3SiBr$$

A 1-L, two- or three-necked flask equipped with a nitrogen inlet, addition funnel, and magnetic stir bar is charged with $(Me_3Si)_2NPMe_2$ (75.6 g, 0.34 mol) and benzene (400 mL, distilled from CaH_2). A solution of Br_2 (57.4 g, 0.36 mol) in benzene (200 mL) is added dropwise to the phosphine (phosphinous amide) solution at $0°$. Addition of Br_2 is stopped when the solution in the flask is slightly yellow due to the presence of excess Br_2. This mixture is stirred at room temperature for 0.5 hr and more Br_2 is added if the yellow color in the solution disappears. Then solvent is removed under reduced pressure and the residue is distilled to give $Me_3SiN{=}P(Br)Me_2$ (bp 55°/5.0 torr) as a very moisture-sensitive liquid that fumes in air and ignites paper. Yield: 69.2 g, 89%. ^{31}P NMR $\delta = 6.82$ (C_6H_6).[4]

3. 2,2,2-Trifluoroethyl *P,P*-Dimethyl-*N*-(trimethylsilyl)phosphinimidate

$$\underset{\overset{|}{Me_3SiN=PMe_2}}{\overset{Br}{}} + CF_3CH_2OH \xrightarrow[-Et_3NHBr]{Et_3N} \underset{\overset{|}{Me_3SiN=PMe_2}}{\overset{OCH_2CF_3}{}}$$

A 1-L, three-necked, round-bottom flask equipped with a nitrogen inlet, mechanical stirrer, and addition funnel is charged with $Me_3SiN=P(Br)Me_2$ (69.2 g, 0.30 mol), benzene (400 mL), and Et_3N (42.3 mL, 0.30 mol, distilled from CaH_2). After the mixture is cooled to 0°, CF_3CH_2OH (22.1 mL, 0.30 mol, distilled from barium oxide)[7] is added over 0.5 hr. This reaction mixture is allowed to warm to room temperature, stirred for 16 to 18 hr, and then filtered under nitrogen. The solids are washed with hexane (3 × 100 mL) and solvent is removed under reduced pressure. The residue is distilled to give 47.6 g (64%) of $Me_3SiN=P(OCH_2CF_3)Me_2$ (bp 54°/11 torr). ^{31}P NMR δ = 32.32 (CDCl$_3$).[4] This slightly moisture-sensitive liquid must be stored at ~ −10° in order to prevent thermal decomposition.

4. Poly(dimethylphosphazene)

$$\underset{\overset{|}{Me_3SiN=PMe_2}}{\overset{OCH_2CF_3}{}} \xrightarrow{\Delta} \left(\!\!-N=\overset{\overset{\displaystyle Me}{|}}{\underset{\underset{\displaystyle Me}{|}}{P}}\!\!-\right)_{\!n} + Me_3SiOCH_2CF_3$$

A heavy-walled glass ampule (~20-mL capacity) with a constriction is purged with nitrogen and charged with $Me_3SiN=P(OCH_2CF_3)Me_2$ (6.46 g, 26.1 mmol). After degassing by the freeze–pump–thaw method, the ampule is sealed with a torch at the constriction and placed in either an oil bath at 189° for 41 hr (*behind a safety shield*) or in an oven at 160° for 65 hr. The ampule is opened and the volatile component ($Me_3SiOCH_2CF_3$) is trapped on a vacuum line (4.45 g, 99.0%). The remaining white solid is dissolved in CH_2Cl_2 and precipitated by pouring the solution into hexane. Drying under vacuum yields 1.93 g (99%) of $(NPMe_2)_n$.

Anal. Calcd. for C_2H_6PN: C, 32.01; H, 8.06; N, 18.66. Found: C, 31.75; H, 8.21; N, 18.32.

Larger quantities of polymer can be prepared by carrying out the thermolysis in a stainless steel reaction bomb fitted with a valve for removal of the volatile by-product. In this manner 31.4 g (0.126 mol) of $Me_3SiN=P(OCH_2CF_3)Me_2$ was heated at 160° for 65 hr or 127 g (0.51 mol) for 302 hr to give complete thermolysis to $(NPMe_2)_n$.

Properties

Poly(dimethylphosphazene) is an air-stable, white, film-forming polymer that is soluble in CH_2Cl_2, $CHCl_3$, EtOH, and $THF-H_2O$ (50:50), and is insoluble in H_2O, acetone, THF, and Et_2O. Its melting point is 148–149° and the glass transition temperature is $-40°$.[5] Molecular weight (M_w) has been determined by light scattering to be ~50,000. Intrinsic viscosities of various samples range from 41 to 80 mL/g (CH_2Cl_2). The ^{31}P NMR spectrum is a singlet at 8.26 ppm ($CDCl_3$) and the ^1H and ^{13}C NMR spectra have broad doublets at 1.43 ppm (J_{PH} = 12.5 Hz, in CH_2Cl_2) and 22.46 ppm (J_{PC} = 90.2 Hz, in $CDCl_3$).

B. POLY(METHYLPHENYLPHOSPHAZENE)

Procedure[8]

1. *P*-Methyl-*P*-phenyl-*N*,*N*-bis(trimethylsilyl)phosphinous Amide

$$(Me_3Si)_2NH \xrightarrow[\text{(2) PhPCl}_2]{\text{(1) } n\text{-BuLi}} (Me_3Si)_2NP(Ph)Cl$$

$$(Me_3Si)_2NP(Ph)Cl \xrightarrow{\text{MeMgBr}} (Me_3Si)_2NP(Ph)Me$$

As described in Section A.1, $(Me_3Si)_2NH$ (208.5 mL, 1.0 mol) in 750 mL of dry Et_2O is treated at 0° with n-BuLi (1.0 mol, 1.6 M in hexane). The resulting solution is cooled to $-78°$ and $PhPCl_2$ (135.7 mL, 1.0 mol) is added over ~10 min. After stirring for 1 hr at ambient temperature, MeMgBr (1.0 mol, 3.0 M in Et_2O) is added at 0° over 1.5 to 2 hr. Work up consists of addition of hexane (300 mL), filtration, and washing of solids with hexane (3 × 150 mL); solvent removal from the filtrate at reduced pressure; and distillation of the residue to afford $(Me_3Si)_2NP(Ph)Me$, bp 78°/0.02 torr. Yield: 229.6 g, 81%. ^{31}P NMR δ = 37.6 ($CDCl_3$).[3]

2. 2,2,2-Trifluoroethyl *P*-Methyl-*P*-phenyl-*N*-(trimethylsilyl)phosphinimidate

$$(Me_3Si)_2NP(Ph)Me \xrightarrow[\text{(2) CF}_3\text{CH}_2\text{OH/Et}_3\text{N}]{\text{(1) Br}_2} Me_3SiN{=}\overset{\displaystyle OCH_2CF_3}{\underset{\displaystyle Me}{P}}{-}Ph$$

A 1-l, three-necked flask equipped with a nitrogen inlet, a mechanical stirrer, and a pressure equalizing addition funnel is charged with

$(Me_3Si)_2NP(Ph)Me$ (116.9 g, 0.41 mol) and dry benzene (300 mL). A solution of Br_2 (69.2 g, 0.3 mol in 150 mL of benzene) is added dropwise to the 0° solution until a faint yellow color persists. After stirring for 0.5 hr, the solvent and Me_3SiBr are removed under reduced pressure. The last traces of Me_3SiBr must be removed so it is necessary to check the ^1H NMR[4] spectrum of the crude product at this point. [The P—Br compound $Me_3SiN=P(Br)(Ph)Me$ (^{31}P NMR δ = 0.08) obtained in this reaction will decompose if distillation is attempted[4]]. Freshly distilled benzene (400 mL) and Et_3N (63.0 mL, 0.45 mol) are then added to the flask. Over a 15- to 20-min period, CF_3CH_2OH (29.9 mL, 0.41 mol) is added to the solution at 0° and the resulting mixture is stirred overnight after warming to room temperature. Filtration, washing of the solids with hexane (3 × 100 mL), solvent removal from the combined filtrate and washings, and distillation affords $Me_3SiN=P(OCH_2CF_3)(Ph)Me$ as a colorless liquid, bp 58°/0.2 torr. Yield: 69.3 g, 55%. (*Note:* Yields may vary from 50 to 85%.) ^{31}P NMR δ = 22.24 (CH_2Cl_2).

3. Poly(methylphenylphosphazene)

A stainless steel bomb fitted with a needle valve is purged with nitrogen and charged with $Me_3SiN=P(OCH_2CF_3)(Ph)Me$ (62.0 g, 0.20 mol). After degassing by evacuating for 10 min at room temperature, the bomb is placed in an oven at 190° for 10 days. The bomb is then attached to a vacuum line and the volatile component is transferred to a preweighed flask. If at least a 90% yield (0.18 mol) of $Me_3SiOCH_2CF_3$ is not obtained, the bomb and the remaining contents are returned to the oven for another day. The volatile material is again removed and the process is repeated until a 90 to 100% yield of $Me_3SiOCH_2CF_3$ is obtained. Then the bomb is opened to the atmosphere and $[N=P(Ph)(Me)]_n$ is removed by dissolving in CH_2Cl_2. Pouring this solution into hexane results in precipitation of the polymer.

Anal. Calcd. for C_7H_8PN: 61.32; H, 5.88. Found: C, 61.58; H, 6.13.

Properties

Poly(methylphenylphosphazene) is soluble in CH_2Cl_2, $CHCl_3$, benzene, and THF, and is insoluble in H_2O, hexane, and acetone. Molecular weight

(M_w) by gel permeation chromatography is ~54,000. The polymer is a white to light brown, air stable, brittle material that is readily plasticized by solvents. The glass transition temperature is 37° and it does not show a melting point transition by DSC analysis. The ^{1}H NMR spectrum shows a complex multiplet in the phenyl region and several overlapping doublets centered at 1.42 ppm (J_{PH} = 12.5 Hz, in CH_2Cl_2). In addition to a multiplet in the phenyl region, the ^{13}C NMR spectrum contains three P—Me doublets centered at 22.29 ppm ($J_{PC} \cong$ 90 Hz, in $CDCl_3$). The ^{31}P NMR spectrum shows a broad singlet at 1.69 ppm ($CDCl_3$).

References

1. H. R. Allcock, *Phorphorus–Nitrogen Compounds,* Academic Press, New York, 1972.
2. (a) H. R. Allcock, D. B. Patterson, and T. L. Evans, *J. Am. Chem. Soc.,* **99,** 6095 (1977).
   (b) H. R. Allcock, and C. T.-W. Chu, *Macromolecules,* **12,** 551 (1979).
3. R. H. Neilson, and P. Wisian-Neilson, *Inorg. Chem.,* **21,** 3568 (1982).
4. P. Wisian-Neilson, and R. H. Neilson, *Inorg. Chem.,* **19,** 1875 (1980).
5. R. H. Neilson, R. Hani, P. Wisian-Neilson, J. J. Meister, A. K. Roy, and G. L. Hagnauer, *Macromolecules,* **20,** 910 (1987).
6. R. H. Neilson, and P. Wisian-Neilson, *Chem. Rev.* **88,** 541 (1988).
7. A better drying procedure for CF_3CH_2OH, recommended by the checkers, is distillation from activated 3-Å molecular sieves into a flask containing 3-Å molecular sieves.
8. Unless described otherwise, the procedures from Section A should be used.

17. PENTACHLORO(VINYLOXY)CYCLOTRI-PHOSPHAZENES AND THEIR POLYMERS
[2,2,4,4,6-Pentachloro-6-(ethenyloxy)-1,3,5,2λ^5,4λ^5,6λ^5-triazatriphosphorine]

Submitted by CHRISTOPHER W. ALLEN*, KOLIKKARA RAMACHANDRAN,† and DOUGLAS E. BROWN*
Checked by W. J. BIRDSALL‡ and J. E. SCHEIRER‡

It has recently been found that the variety of new and useful phosphazene derivatives that are available can be expanded dramatically by incorporation of an organofunctional substituent on the phosphazene ring. The reactive center on the side chain can then serve as the site for further tranformations in syntheses. This approach has been most successfully utilized with olefinic,[1] *p*-lithiophenoxy,[2] and *p*-aminophenoxy[3] phospha-

*Department of Chemistry, University of Vermont, Burlington, VT 05405.
†3M Center, 3M Corporation, St. Paul, MN 55144.
‡Albright College, Reading, PA 19603.

zenes. One such class of organofunctional monomers is the (vinyloxy)cyclophosphazenes that are available from the reaction of the enolate anion of acetaldehyde with halocyclophosphazenes.[4-7] The lithium enolate of acetaldehyde is conveniently obtained by metallation of THF,[8] and the reaction of this material with hexachlorocyclotriphosphazene results in the formation of pentachloro(vinyloxy)cyclotriphosphazene, $N_3P_3Cl_5OC_2H_3$.[4] This process is described in the next section. The conversion of this monomer to polypentachloro(vinyloxy)cyclotriphosphazene is described in Section B.

A. PENTACHLORO(VINYLOXY)CYCLOTRIPHOSPHAZENE
(2-Vinyloxy-2,4,4,6,6-pentachlorocyclotriphosphazatriene; 2,2,4,4,6-Pentachloro-6-(ethenyloxy)-1,3,5,2λ^5,4λ^5,6λ^5-triazaphosphorine)

$$C_4H_8O + n\text{-}C_4H_9Li \longrightarrow LiOC_2H_3 + C_2H_4 + C_4H_{10}$$

$$N_3P_3Cl_6 + LiOC_2H_3 \longrightarrow N_3P_3Cl_5OCH{=}CH_2 + LiCl$$

Procedure

The apparatus shown in Figure 1[9] is fitted with a magnetic stirring bar, a 100-ml pressure-equalizing dropping funnel, and septa on the two side arms. The glassware is assembled hot, the stopcocks are closed, and the system is flushed with nitrogen exiting through the dropping funnel to a mercury bubbler. The nitrogen flow is reduced to a minimal rate and 100 mL of dry THF is placed in the apparatus. Using either a syringe or a double-ended needle, 50 mL of a 1.55 M solution of butyllithium in hexanes (0.078 mol) is placed in the addition funnel and slowly added to the stirred THF. Stirring is continued overnight. A 500-mL three-necked flask is attached to the apparatus in the figure and is charged with 15.0 g (0.043 mol)* of hexachlorocyclotriphosphazene, $N_3P_3Cl_6$,† and a magnetic stirring bar. Septa are fitted to the side arms, and the flask is attached to the appartus containing the lithium enolate. The nitrogen line is removed from the side aim on the apparatus shown in the figure and connected to the flask via a syringe needle. The upper stopcock of the apparatus is opened allowing the system to be flushed with nitrogen. Approximately 100 mL of dry THF is added to the flask, and stirred until the solid dissolves. The phosphazene containing flask is immersed in an ice bath and the enolate

---

*The high enolate–phosphazene ratio is used to ensure reaction of all of the $N_3P_3Cl_6$, which is difficult to separate completely from the desired product.
†Shin Nisso Kako Co., ltd. 3-1.60, Ukima, Kita-ku, Tokyo, Japan 115.

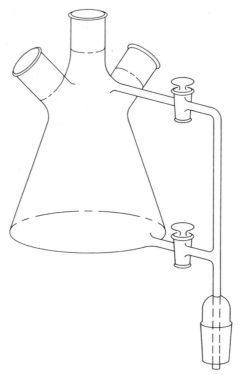

Fig. 1.

solution is slowly added. After complete addition, the stirred reaction mixture is allowed to warm to room temperature. The solvent is removed by means of a rotary evaporator, and the resulting oily mixture is treated with activated charcoal and 200 mL of low boiling (30–60°) petroleum. After filtration through celite or Filter-Aid, the solvent is removed and the process is repeated with 100 mL of petroleum and additional activated charcoal. After removal of the solvent an oily mixture remains. Flash chromatography[10] is an efficient method of separating the monosubstituted derivative from materials with higher degrees of substitution. A 4-cm diameter chromatographic column fitted with a 35/20 ball joint at the top and leading to a nitrogen inlet is filled with 11 cm of flash chromatography grade (~4 μm average particle diameter) silica gel (Baker) in a slurry with low boiling petroleum. A 5.56-g sample of the crude product is placed on the column, solvent is added, the ball joint is clamped, and the nitrogen pressure is adjusted to provide a slow flow through the column. Fractions are collected in test tubes and monitored by TLC. The first component to be eluted is the desired product, and all fractions containing this material are combined. The solvent is removed by means of rotary evaporation, kept at 40°. Purification is effected by short path distillation at a bath temperature of 85° (0.02 torr) to yield 2.25 g of a water white liquid, bp 55 to 57°.

Anal. Calcd. for $N_3P_3Cl_5OC_2H_3$: C, 6.75, H, 0.84; mol wt 353. Found: C, 6.74; H, 0.75; M_w 353 (mass spectrum).

Properties

Pentachloro(vinyloxy)cyclotriphosphazene is a colorless liquid stable to atmospheric hydrolysis. It is soluble in common organic solvents. The IR spectrum exhibits a strong phosphorus–nitrogen ring stretching band at 1220 cm^{-1}.[4] The ^{31}P NMR spectrum shows resonances for $=PCl(OC_2H_3)$ centers at 13.2 ppm and $=PCl_2$ centers at 23.4 ppm with $^2J_{PNP} = 64$ Hz.[4]

B. POLY[PENTACHLORO(VINYLOXY)CYCLOTRI-PHOSPHAZENE] (Poly[2-vinyloxy-2,4,4,6,6-pentachlorocyclotriphosphaztriene]; Poly[2,2,4,4,6-pentachloro-6-(ethenyloxy)-1,3,5,2λ⁵,4λ⁵,6λ⁵-triaztriphosphorine])

$$N_3P_3Cl_5OC_2H_3 \xrightarrow[\Delta]{Me_2C(CN)N=NC(CN)Me_2} (CHCH_2)_n$$
$$\qquad\qquad\qquad\qquad\qquad\qquad\qquad | $$
$$\qquad\qquad\qquad\qquad\qquad\qquad\quad ON_3P_3Cl_5$$

Polymerization through the vinyl function in pentachloro(vinyloxy)-cyclotriphosphazene can be accomplished under radical initiation conditions. The presence of moisture and (vinyloxy)phosphazenes with degrees of substitution >1 must be avoided because the former can effect chain transfer and the latter results in the formation of cross-linked materials.

Procedure

A 2.25-g sample of pentachloro(vinyloxy)cyclotriphosphazene is distilled at 0.02 torr from P_4O_{10} to a flask containing 0.02 g of 2,2^1-azobis(2-meth-ylpropanenitrile) [azobis(isobutyronitrile)] (AIBN).* After distillation, the

---

*AIBN is purified by vacuum sublimation at ambient temperature.

stopcock leading to the vacuum system is closed, and the flask (still attached to the distillation apparatus) is placed in a constant temperature bath at 60°. The flask is occasionally swirled to aid in dissolution of AIBN and to recover AIBN, which condenses on the upper walls of the flask. The transition to a very viscous medium occurs in approximately 2 hr. At this point, the mixture is dissolved in 50 mL of dichloromethane and the resulting solution is dropped slowly down the wall of a 400-mL beaker containing 200 mL of stirred methanol. The polymer is allowed to settle, and most of the solvent is decanted. Approximately 200 mL of additional methanol is added and then decanted after the polymer has settled. The polymer is isolated by filtration through coarse filter paper, dried in an oven at 50°, and then placed *in vacuo* (0.02 torr) overnight. A yield of 0.27 g (12% conversion) is obtained. Conversions of 10 to 20% are typically achieved with higher conversions being possible with extended reaction times.

Anal. Calcd. for $N_3P_3Cl_5OC_2H_3$: C, 6.75; H, 0.84. Found: C, 7.16; H, 0.93.

Properties

Poly[pentachloro(vinyloxy)cyclotriphosphazene] is a white solid that is stable to atmospheric hydrolysis. It is soluble in toluene, dichloromethane, and so on, and can be cast into flexible thin films. Molecular weights will vary with experimental conditions with M_n as high as 2.85×10^5 (membrane osmometry) being observed. Thermal decomposition of the polymer is a complex process, with the first stage being elimination of HCl starting around 120°. Nucleophilic substitution reactions on the cyclophosphazene groups in the polymer allow for the synthesis of a broad range of related polymers.

References

1. C. W. Allen, *J. Polym. Sci. Polym. Symp.*, **70**, 79 (1983).
2. H. R. Allcock, T. L. Evans, and T. J. Fuller, *Inorg. Chem.*, **19**, 1026 (1980).
3. H. R. Allcock, P. E. Austin, and T. R. Rakowsky, *Macromolecules*, **14**, 1622 (1981); H. R. Allcock, W. C. Hymer, and P. E. Austin, *ibid.* **16**, 1401 (1983).
4. C. W. Allen, K. Ramachandran, R. P. Bright, and J. C. Shaw, *Inorg. Chim. Acta*, **64**, L109 (1982).
5. K. Ramachandran and C. W. Allen, *Inorg. Chem.*, **22**, 1445 (1983).
6. P. J. Harris, M. A. Schwalke, V. Liu, and B. L. Fisher, *Inorg. Chem.*, **22**, 1812 (1983).
7. C. W. Allen and R. P. Bright, *Inorg. Chim. Acta*, **99**, 107 (1985).
8. R. B. Bates, L. M. Kroposki, and L. M. Potter, *J. Org. Chem.*, **37**, 560 (1972).
9. C. W. Allen, R. P. Bright, J. L. Desorcie, J. A. MacKay, and K. Ramachandran, *J. Chem. Educ.*, **57**, 564 (1980).
10. W. C. Still, M. Kahn, and A. Mitra, *J. Org. Chem.*, **43**, 2923 (1978).

Chapter Three

COMPOUNDS OF
PHARMACOLOGICAL INTEREST

18.   BORON ANALOGS OF AMINO ACIDS

Submitted by BERNARD F. SPIELVOGEL,* FAHIM U. AHMED,* and
ANDREW T. McPHAIL*
Checked by KAREN W. MORSE† and TERRY J. LOFTHOUSE†

Amine carboxyboranes such as $H_3NBH_2CO_2H$ (ref. 1) and $Me_3NBH_2CO_2H$ (ref. 2) are isoelectronic and isosteric (protonated) boron analogs of the α-amino acids glycine, $H_3\overset{+}{N}CH_2CO_2^-$, and betaine, $Me_3\overset{+}{N}CH_2CO_2^-$, respectively. Considerable interest may be expected in these boron analogs in view of the profound biological activity of the α-amino acids. Indeed, these and other examples of boron analogs of the α-amino acids, together with their precursors and derivatives have been found to possess interesting pharmacological activity including significant hypolipidemic, antiarthritic, and antitumor activity[3-5] in animal model screens.

Procedures are given for the preparation of trimethylamine-cyano borane, Me_3NBH_2CN, the precursor to trimethylamine-carboxyborane, $Me_3NBH_2CO_2H$, as well as for $Me_3NBH_2CO_2H$ and its N-ethylamide and

*Paul M. Gross Chemical Laboratory, Duke University, Durham, NC 27706.
†Departments of Chemistry and Biochemistry, Utah State University, Logan, UT 84322.

methyl ester derivatives. These four compounds can serve as the basis for the synthesis of many other interesting derivatives including peptides.

A. TRIMETHYLAMINE-CYANOBORANE

$$NaBH_3CN + Me_3N \cdot HCl \xrightarrow[\Delta]{THF} Me_3NBH_2CN + NaCl + H_2$$

Trimethylamine-cyanoborane is prepared by reacting trimethylamine-hydrochloride and $NaBH_3CN$ in refluxing THF.[6] It also has been prepared by the addition of Me_3N to independently generated species or adducts containing $—BH_2CN$,[7,8] by the interaction of Me_3NBH_2I and NaCN in ether solvents,[9] and other routes.[10] The present procedure, however, is a very convenient one step, high yield synthesis using commercially available starting materials.

Procedure

■ **Caution.** *NaBH₃CN is a toxic chemical. Considerable care should be exercised not to inhale the powder or to contact the chemical or its solutions. Rubber gloves should be worn and all operations carried out in an efficient hood. Me₃NBH₂CN is also toxic (see Properties) and similar care should be taken in working with the compound. Although Me₃NBH₂CN has shown significant pharmacological activity at doses below toxic levels in animal model studies,[4-6] very little toxicological data is available. It should be used only for experimental laboratory use and is not to be used as a drug on humans.*

A 2-L three-necked flask is equipped with a nitrogen inlet, reflux condenser (connected to an oil bubbler to monitor H_2 gas evolution) and magnetic stirrer. For more efficient stirring, a mechanical stirrer is better. After flushing with nitrogen, 75.40 g (1.20 mol) of NaBH₃CN* is added followed by 800 mL of dry THF. The mixture is stirred until most of the NaBH₃CN is solubilized. To the stirring solution, 128.06 g (1.34 mol, ~11% excess) of Me₃N·HCl is added slowly followed by an additional 200 mL of dry THF. After the H_2 gas evolution has slowed considerably, the suspension is refluxed for 48 hr. During this period, considerable amounts of foam are produced. The reaction mixture is then cooled and filtered, the solid (NaCl) is washed with THF (2 × 50 mL), and the solvent removed from the filtrate at reduced pressure leaving 104.6 g (89.1%) of crude

---

*Commercially available from Aldrich Chemical Company and Alpha Inorganics. NaBH₃CN is quite hygroscopic and is preferably weighed out in a dry atmosphere.

Me$_3$NBH$_2$CN, mp 60°. The crude product is readily purified by recrystallization from THF–petroleum or by sublimation under reduced pressure to give an 82% yield (mp 63°).*

Anal. Calcd. for (C$_6$H$_{17}$BN$_2$O): C,H,B,N. Analyses for indicated elements within ±0.3% were obtained.

Properties

Trimethylamine-cyanoborane is a white crystalline slightly hygroscopic solid, soluble in most organic solvents; mp 63°: IR (Nujol) v_{BH}(2400), v_{CN}(2200) cm^{-1}; ^{1}H NMR (CDCl$_3$) δ = 2.74 (s, CH$_3$N); ^{11}B NMR (CDCl$_3$, BF$_3$·Et$_2$O std) δ = −13.16 (t, J_{B-H} = 105 Hz). Its purity can be checked by melting point and ^{1}H NMR. The compound is quite stable towards hydrolysis. It is toxic with an LD$_{50}$ of 70 mg/kg in male mice[3] although it possesses significant pharmacological activity at doses considerably below this value.[3-5]

B. TRIMETHYLAMINE-CARBOXYBORANE (BORON ANALOG OF BETAINE)

$$Me_3NBH_2CN + Et_3O^+BF_4^- \xrightarrow[\Delta]{CH_2Cl_2} Me_3NBH_2CNEt^+BF_4^- + Et_2O$$

$$Me_3NBH_2CNEt^+BF_4^- + 2H_2O \longrightarrow Me_3NBH_2CO_2H + EtNH_3^+BF_4^-$$

Trimethylamine-carboxyborane is obtained[2] by activating the cyano group in Me$_3$NBH$_2$CN by ethylation with triethyloxonium tetrafluoroborate, followed by hydrolysis of the resulting nitrilium salt with water. Attempts to prepare Me$_3$NBH$_2$CO$_2$H from Me$_3$NBH$_2$CN directly by acid, base hydrolysis, or KMnO$_4$ oxidation–hydrolysis under various conditions have been unsuccessful resulting instead of B—H hydrolysis (under forcing conditions) to give boric acid or borate salts. Freshly prepared Et$_3$O$^+$BF$_4^-$ (ref. 11) always gives better yields of Me$_3$NBH$_2$CO$_2$H than the commercially available material. In some instances, no product is obtained in using commercial sources of Et$_3$O$^+$BF$_4^-$, which may reflect the instability of the compound and its solutions and prior handling.

---

*The checkers reported a crude yield of 89.9% and that some BH$_3$CN$^-$ (^{11}B NMR) remained after the first recrystallization. A pure product (^{11}B NMR) was obtained after the second recrystallization with a yield of 65%.

Procedure

■ **Caution.** *Me$_3$NBH$_2$CN is a toxic compound (see Properties) and proper care should be exercised in handling it. Although Me$_3$NBH$_2$CO$_2$H is a relatively non toxic compound (see Properties) and has been found to possess significant pharmacological activities in animal model studies, very little toxicological data is available; it should be used only for experimental laboratory use and is not to be used as a drug on humans.*

To a 500-mL three-necked flask equipped with reflux condenser, magnetic stirrer, and N$_2$ inlet, is transferred 32.53 g (0.28 mol) of Me$_3$NBH$_2$CN and 240 mL of 2 M Et$_3$OBF$_4$ (0.48 mol, ~100% excess) in CH$_2$Cl$_2$ under nitrogen. The resulting clear solution is refluxed (45°) for 24 hr. The solution is then cooled and the solvent removed at reduced pressure leaving a white solid material that is vacuum pumped overnight to remove BF$_3$·Et$_2$O formed from decomposition of some Et$_3$O$^+$BF$_4^-$ during the period of reflux. To this solid (*N*-ethylnitrilium salt, Me$_3$NBH$_2$CNEt$^+$BF$_4^-$), 200 mL of water is added and the solution is stirred at room temperature for 2½ days. The aqueous solution (some solid product forms at this stage) is repeatedly extracted with CH$_2$Cl$_2$ (4 × 100 mL). The organic extracts are dried over MgSO$_4$, filtered and evaporated to dryness. The crude solid Me$_3$NBH$_2$COOH is purified by recrystallization from warm water (at least 55°) giving 21.38 g (65.8%) white plates, mp 147° (dec, under vacuum).

If freshly prepared Et$_3$O$^+$BF$_4^-$ is used, an excess of 30% is found to give a similar yield of Me$_3$NBH$_2$CO$_2$H.

Anal. Calcd. (C$_4$H$_{12}$BNO$_2$) C,H,B,N: Analyses for indicated elements within ±0.3% were obtained.

Properties

The product is a white crystalline solid, soluble in H$_2$O and common organic solvents. It melts with decomposition at 147°. It is stable in H$_2$O and air but will decompose approximately 25% in 1 *N* HCl after 1 week. The IR (KBr) ν_{OH}(3130), ν_{BH}(2380), ν_{CO}(1645) cm^{-1}. ^{1}H NMR (D$_2$O) δ = 2.72 (s, Me$_3$N), 4.65 (s, HDO); ^{11}B NMR (CDCl$_3$, BF$_3$·Et$_2$O std): δ = −9.7 (t, J_{B-H} = 98 Hz). The purity of Me$_3$NBH$_2$CO$_2$H is most easily checked by ^{11}B NMR since its chemical shift (−9.7 ppm) is 3.5-ppm downfield from that of Me$_3$NBH$_2$CN, the major impurity. Trimethylamine-carboxyborane is a relatively nontoxic compound having an LD$_{50}$ of 1800 mg/kg in mice.[3]

The amine group of Me$_3$NBH$_2$CO$_2$H can be exchanged with other amines, for example, Me$_2$NH,[12] MeNH$_2$,[12] and NH$_3$,[1] the last giving H$_3$NBH$_2$CO$_2$H, the boron analog of glycine. Me$_3$NBH$_2$CO$_2$H and other amine-carboxyboranes are very weak acids[13] with pK_a for ionization of the carboxyl proton being around 8, some sixfold higher in pK than the analogous amino acid. $Me_3NBH_2CO_2H$ forms a dinuclear complex with Cu(II) similar to dinuclear copper acetate complexes. The complex has enhanced antitumor and hypolipidemic activity.[14,15]

C. TRIMETHYLAMINE-(ETHYLCARBAMOYL)BORANE

$$Me_3NBH_2CN + Et_3O^+BF_4^- \xrightarrow[\Delta]{CH_2Cl_2} Me_3NBH_2CNEt^+BF_4^- + Et_2O$$

$$Me_3NBH_2CNEt^+BF_4^- + NaOH \longrightarrow Me_3NBH_2C(O)NHEt + NaBF_4$$

The N-ethylamide of $Me_3NBH_2CO_2H$, $Me_3NBH_2C(O)NHEt$, or trimethylamine (ethylcarbamoyl)borane is prepared,[1,16] by ethylating Me_3NBH_2CN with $Et_3O^+BF_4^-$ in refluxing CH_2Cl_2 (45°) followed by rapid hydrolysis of the resulting nitrilium salt with 1N NaOH at 0°.

Procedure

■ **Caution.** *Me_3NBH_2CN is a toxic compound (see Properties). $Me_3NBH_2C(O)NHEt$ should be used only for experimental laboratory use and is not to be used as a drug on humans.*

A solution of Me_3NBH_2CN (11.9 g, 0.12 mol) and 250 mL of 1 M Et_3OBF_4 (ref. 11) in CH_2Cl_2 (0.25 mol) was refluxed (45°) under N_2 for 24 hr. The reaction mixture is cooled to 0°, and 1 N NaOH is added slowly with vigorous stirring until the solution is quite basic (pH > 13). After stirring (maintaining pH > 13) for 1 hr at room temperature, the organic layer is separated and the aqueous layer extracted three times with CH_2Cl_2. The organic portions are combined and dried over $MgSO_4$, and the solvent removed *in vacuo*. The remaining viscous liquid is distilled under vacuum with minimum heating (oil bath) to give 13.1 g (75%) of oily amide; bp 80° (0.15 torr).

Anal. Calc. for $C_6H_{17}BN_2O$: C, 50.10; H, 11.83; N, 19.49; B, 7.45. Found: C, 49.86; H, 11.69; N, 19.59, B, 7.50

Properties

Trimethylamine-(ethylcarbamoyl)borane is a viscous malodorous liquid and undergoes considerable decomposition at temperatures much above 80°. IR (neat): ν_{NH} (3289), ν_{BH} (2330), $\nu_{C(O)NH}$ (1590, 1480) cm^{-1}; ^1H NMR

(CDCl$_3$): δ = 1.07 (t, CH$_3$CH$_2$), δ = 2.75 (s, Me$_3$N), δ = 3.57 (m, CH$_2$CH$_3$), δ = 5.43 (br. s, NH); ^{11}B NMR (CDCl$_3$, BF$_3$·Et$_2$O std): δ = -7.4 (t, J_{BH} = 90Hz). Its purity can be checked by ^{1}H and ^{11}B NMR spectral data, which are different from the Me$_3$NBH$_2$CO$_2$H and Me$_3$NBH$_2$CN impurities. Bases such as Me$_2$NH, MeNH$_2$, and NH$_3$ in excess can displace Me$_3$N in Me$_3$NBH$_2$C(O)NHEt to give the corresponding (ethylcarbamoyl)boranes.[16] Me$_3$NBH$_2$C(O)NHEt is considerably basic and reacts readily with anhydrous HCl to form a novel protonated salt,[17] Me$_3$NBH$_2$C(OH)=NHEt$^+$ Cl$^-$. Trimethylamine-(ethylcarbamoyl)borane has an LD$_{50}$ of 320 mg/kg in mice.[3]

D. TRIMETHYLAMINE-CARBOMETHOXYBORANE

$$Me_3NBH_2CO_2H + Cl\overset{\overset{\displaystyle O}{\|}}{C}OCH_3 + Et_3N \xrightarrow[\text{cat, } 0°]{CH_2Cl_2}$$

$$Me_3NBH_2C(O)OMe + Et_3N\cdot HCl + CO_2$$

Trimethylamine-carbomethoxyborane is prepared[18] in 84% yield by reacting Me$_3$NBH$_2$CO$_2$H with methyl chloroformate for 1 hr at 0°. This ester has also been prepared in 82% yield by condensing Me$_3$NBH$_2$CO$_2$H and CH$_3$OH with dicyclohexylcarbodiimide at ambient temperature for 1 week.[19]

Procedure

The following procedure should be carried out under N$_2$. To a solution of Me$_3$NBH$_2$COOH (1.17 g, 0.01 mol) and Et$_3$N (1.11 g, 0.011 mol) in CH$_2$Cl$_2$ (100 mL) at 0° is added methyl chloroformate (0.945 g, 0.01 mol), followed by dimethylaminopyridine (0.122 g, 0.001 mol). The resulting solution is stirred at 0° for 1 hr, washed with water (2 × 20 mL), dried over MgSO$_4$ and concentrated to give pure ester; Yield: 1.1 g (84%).*

Anal. Calcd. for C$_5$H$_{14}$BNO$_2$: C, 45.85; H, 10.77; N, 10.69. Found: C, 45.95; H, 10.99; N, 10.56%.

Properties

Trimethylamine-carbomethoxyborane is a sweet smelling solid. It can be purified by recrystallization (CH$_2$Cl$_2$–pentane) or sublimation at reduced

*The checkers reported yields of 62 and 66% in two attempts, mp 90 to 92°.

pressure; IR (CDCl$_3$): ν_{BH} (2385), ν_{CO} (1660) cm^{-1}; ^{1}H NMR (CDCl$_3$): δ = 2.72 (s, Me$_3$N), δ = 3.48 (s, OCH$_3$); ^{11}B NMR (CDCl$_3$, BF$_3$·Et$_2$O): δ = -9.09 (t, J_{BH} = 99 Hz). Its purity can be determined by mp and ^{1}H NMR spectral data. H$_3$NBH$_2$CO$_2$Me can be prepared from Me$_3$NBH$_2$CO$_2$Me by reacting it with NH$_3$ in a displacement reaction.[19] Me$_3$NBH$_2$C(O)OMe is a potent hypolipidemic agent in animal model screens.[20]

References

1. B. F. Spielvogel, M. K. Das, A. T. McPhail, K. D. Onan, and I. H. Hall, *J. Am. Chem. Soc.*, **102**, 6343 (1980).
2. B. F. Spielvogel, L. Wojnowich, M. K. Das, A. T. McPhail, and K. D. Hargrave, *J. Am. Chem. Soc.*, **98**, 5702 (1976).
3. I. H. Hall, C. O. Starnes, B. F. Spielvogel, P. Wisian-Neilson, M. K. Das, and L. Wojnowich, *J. Pharm. Sci.*, **68**, 685 (1979).
4. I. H. Hall, C. O. Starnes, A. T. McPhail, P. Wisian-Neilson, M. K. Das, F. Harchelroad, Jr., and B. F. Spielvogel, *J. Pharm. Sci.*, **69**, 1024 (1980).
5. (a) I. H. Hall, M. K. Das, F. Harchelroad, Jr., P. Wisian-Neilson, A. T. McPhail, and B. F. Spielvogel, *J. Pharm. Sci.*, **70**, 339 (1981). (b) I. H. Hall, C. J. Gilbert, A. T. McPhail, K. W. Morse, K. Hassett, and B. F. Spielvogel, *J. Pharm. Sci.*, **74**, 755 (1985).
6. P. Wisian-Neilson, M. K. Das, and B. F. Spielvogel, *Inorg. Chem.*, **17**, 2327 (1978).
7. (a) S. S. Uppal and H. C. Kelly, *Chem. Commun.*, **1970**, 1619. (b) C. Weidig, S. S. Upal, and H. C. Kelly, *Inorg. Chem.*, **13**, 1763 (1974).
8. D. R. Martin, M. A. Chiusano, M. L. Denniston, D. J. Dye, E. D. Martin, and B. T. Pennington, *J. Inorg. Nucl. Chem.*, **40**, 9 (1978).
9. P. J. Bratt, M. P. Brown, and K. R. Seddon, *J. Chem. Soc. Dalton Trans.*, **1974**, 2161.
10. O. T. Beachley and B. Washburn, *Inorg. Chem.*, **14**, 120 (1975).
11. H. Meerwein, *Org. Synth.*, **46**, 113 (1966).
12. B. F. Spielvogel, *Boron Chemistry*, IUPAC, Inorganic Chemistry Division, R. W. Parry and G. Kodoma (eds.), Pergamon, New York, 1980, pp. 119–129.
13. K. H. Scheller, R. B. Martin, B. F. Spielvogel, and A. T. McPhail, *Inorg. Chim. Acta*, **57**, 227 (1982).
14. I. H. Hall, W. L. Williams, C. J. Gilbert, A. T. McPhail, and B. F. Spielvogel, *J. Pharm. Sci.*, **73**, 973 (1984).
15. I. H. Hall, B. F. Spielvogel, and A. T. McPhail, *J. Pharm. Sci.*, **73**, 222 (1984).
16. B. F. Spielvogel, F. U. Ahmed, K. W. Morse, and A. T. McPhail, *Inorg. Chem.*, **23**, 1776 (1984).
17. B. F. Spielvogel, F. U. Ahmed, and A. T. McPhail, *Abstracts of Papers*, 187th National Meeting of the American Chemical Society, St. Louis, MO, April 8–13, 1984; American Chemical Society, Washington, DC; Paper No. INOR 240.
18. B. F. Spielvogel, F. U. Ahmed, and A. T. McPhail, *Synthesis*, **1986**, 833.
19. B. F. Spielvogel, F. U. Ahmed, G. L. Silvey, P. Wisian-Neilson, and A. T. McPhail, *Inorg. Chem.*, **23**, 4322 (1984).
20. I. H. Hall, B. F. Spielvogel, A. Sood, F. Ahmed, and S. Jafri, *J. Pharm. Sci.*, **76**, 359 (1987).

19. 1-AZIRIDINYL-AMINO SUBSTITUTED CYCLOPHOSPHAZENES

Submitted by J. C. VAN DE GRAMPEL,* A. A. VAN DER HUIZEN,*
N. H. MULDER,† J. W. RUSCH,* and T. WILTING*
Checked by M. A. BAXTER‡ and D. P. MACK‡

It has been shown that 1-aziridinyl derivatives of $(NPCl_2)_3$ and $(NPCl_2)_4$ possess *in vitro* and *in vivo* cytostatic activity, the degree depending on the geometrical arrangements of the 1-aziridinyl groupings and on the nature of the other substituents.[1-5] In general the aziridinyl-amino substituted cyclophosphazenes exhibit the highest activity both *in vitro* and *in vivo*. Moreover, for compounds $N_3P_3Az_2(NHMe)_4$ and $N_4P_4Az_2(NHMe)_6$ (Az = 1-aziridinyl) it has been demonstrated that the nongeminal isomers are more active than the geminal ones.[3,5]

Aziridinyl-amino derivatives of $(NPCl_2)_3$ and $(NPCl_2)_4$ can be prepared by the aminolysis of the appropriate 1-aziridinyl-chloro precursors. An alternative possibility is offered by the reaction of compounds $N_3P_3Am_{6-n}Cl_n$ or $N_4P_4Am_{8-n}Cl_n$ (Am = amino) with an excess of aziridine. Both routes provide comparable yields. The choice between them is governed by the accessibility of the precursor.

A. BIS(AZIRIDINYL)TETRAKIS(METHYLAMINO)-CYCLOTRIPHOSPHAZENE [*cis-* AND *trans*-2,4,-BIS(1-AZIRIDINYL)-2,4,6,6-TETRAKIS(METHYLAMINO)-1,3,5,2λ^5,4λ^5,6λ^5-TRIAZATRIPHOSPHORINE AND 2,2-BIS(1-AZIRIDINYL)-4,4,6,6-TETRAKIS(METHYLAMINO)-1,3,5,2λ^5,4λ^5,6λ^5-TRIAZATRIPHOSPHORINE]

$$N_3P_3Cl_6 \xrightarrow{(CH_2)_2NH} N_3P_3AzCl_5 \ + \ \ldots \ +$$

trans-2,4-$N_3P_3Az_2Cl_4$

*Department of Inorganic Chemistry, University of Groningen, Nijenborgh 16, 9747 AG Groningen, The Netherlands.

†Division of Medical Oncology, Department of Internal Medicine, University Hospital, Oostersingel 59, 9713 EZ Groningen, The Netherlands.

‡Department of Chemistry, Shippensburg University, Shippensburg, PA, 17257.

$$2,2\text{-}N_3P_3Az_2Cl_4 \qquad\qquad cis\text{-}2,4\text{-}N_3P_3Az_2Cl_4$$

$$+ \quad 2,2,4\text{-}N_3P_3Az_3Cl_3 \quad + \quad (CH_2)_2NH \cdot HCl \tag{1}$$

$cis\text{-}, trans\text{-}2,4\text{-}N_3P_3Az_2Cl_4$

$$\text{or } 2,2\text{-}N_3P_3Az_2Cl_4 + 8MeNH_2 \longrightarrow cis\text{-}, trans\text{-}2,4\text{-}N_3P_3Az_2(NHMe)_4$$

$$\text{or } 2,2\text{-}N_3P_3Az_2(NHMe)_4 + 4MeNH_2 \cdot HCl \tag{2}$$

Procedure

■ **Caution.** *Aziridine and benzene are carcinogenic and must be handled in an efficiently ventilated hood. Compounds $N_3P_3Az_{6-n}R_n$ have to be considered as toxic; therefore, direct contact of these solids with the skin must be avoided. Care must also be exercized when cleaning the glassware. Diluted hydrochloric acid is recommended for destroying residual aziridine and aziridinyl derivatives.*

The reactions should be carried out in a dry nitrogen atmosphere in order to avoid hydrolysis of partially aminolyzed products.

1. Compounds $N_3P_3Az_{6-n}Cl_n$

A Schlenk flask (200 mL), fitted with a Teflon-coated magnetic stirring bar and a pressure-equalizing dropping funnel topped with a rubber septum, is charged with 10.40 g (29.9 mmol) of $(NPCl_2)_3$ (ref. 6) and 100 mL of benzene, dried over molecular sieves (4 Å). Via a syringe the dropping funnel is charged through the septum with 6.2 mL (120 mmol) of aziridine* (distilled from KOH pellets) and 50 mL of benzene. Under vigorous stirring, the aziridine solution is added dropwise to the solution in the flask, cooled at 6°. After addition, the reaction mixture is slowly warmed to room temperature and stirred during an additional period of 18 hr. Precipitated (polymeric) aziridinium chloride salts are removed by filtration. After thorough washing of the precipitate with benzene, the combined filtrates are

---

*If aziridine cannot be obtained commercially it can be prepared following Wenker's method[7] as modified by Leighton et al.[8] and Reeves et al.[9]

evaporated to dryness *in vacuo*. The resulting crude product, a white, waxy material, is freed from residual polymeric products by chromatography (Silica Woelm 32-63 column, O.D. 20 mm, length 30 cm) applying an hexane–diethyl ether (3:2) mixture as eluent. The separation of the reaction products is performed by HPLC on a Lichrosorb Si 60/10 column (Chrompack, O.D. 22 mm, length 30 cm) using an hexane–diethyl ether (2:1) mixture as eluent.

Fraction 1. 1.50 g of a white solid; recrystallization from hexane yields 1.06 g (2.99 mmol = 10.0%) of $N_3P_3AzCl_5$, mp 67–68.5°.

Anal. Calcd. for $C_2H_4N_4P_3Cl_5$: C, 6.78; H, 1.14; N, 15.82; Cl, 50.03. Found: C, 6.93; H, 1.12; N, 15.61; Cl, 49.60.

Fraction 2. 1.87 g of a colorless oil; recrystallization from hexane yields 1.49 g (4.13 mmol = 13.8%) of *trans*-2,4-$N_3P_3Az_2Cl_4$, mp 66.5–68°.

Anal. Calcd. for $C_4H_8N_5P_3Cl_4$: C, 13.31; H, 2.24; N, 19.41; Cl, 39.29. Found: C, 13.27; H, 2.25; N, 19.53; Cl, 39.01.

Fraction 3. 1.86 g of a white solid; recrystallization from an hexane–diethyl ether mixture yields 1.43 g (3.96 mmol = 13.3%) of 2,2-$N_3P_3Az_2Cl_4$, mp 105.5–107°.

Anal. Calcd. for $C_4H_8N_5P_3Cl_4$: C, 13.31; H, 2.24; N, 19.41; Cl, 39.29. Found: C, 13.40; H, 2.11; N, 19.30; Cl, 39.53.

Fraction 4. 1.48 g of a colorless oil; recrystallization from hexane yields 1.06 g (2.94 mmol = 9.8%) of *cis*-2,4-$N_3P_3Az_2Cl_4$, mp 65.5–67°.

Anal. Calcd. for $C_4H_8N_5P_3Cl_4$: C, 13.31; H, 2.24; N, 19.41; Cl, 39.29. Found: C, 13.35; H, 2.29; N, 19.18; Cl, 39.54.

Fraction 5. 0.84 g of a colorless oil; recrystallization from an hexane–diethyl ether mixture yields 0.63 g (1.71 mmol = 5.7%) of 2,2,4-$N_3P_3Az_3Cl_3$, mp 61.5–63°.

Anal. Calcd. for $C_6H_{12}N_6P_3Cl_3$: C, 19.61; H, 3.29; N, 22.87; Cl, 28.92. Found: C, 19.59; H, 3.20; N, 22.74; Cl, 28.72.

Properties

Compounds $N_3P_3AzCl_5$, $N_3P_3Az_2Cl_4$, and $2,2,4\text{-}N_3P_3Az_3Cl_3$ are white, crystalline solids. They are slightly sensitive to hydrolysis and should be kept in a dry atmosphere at low temperature (about 0°). ^{31}P $\{^1H\}$ NMR data [80.9 MHz; $CDCl_3$; $(NPCl_2)_3$ solution in $CDCl_3$, 19.9 ppm, as external reference] for the bis(1-aziridinyl) derivatives, including those for $N_3P_3AzCl_5$ and $2,2,4\text{-}N_3P_3Az_3Cl_3$, are given in Table I.

The crystal structure of *trans*-$2,4\text{-}N_3P_3Az_2Cl_4$ has been published.[10]

2. Compounds $N_3P_3Az_2(NHMe)_4$

A Schlenk flask (100 mL), equipped with a Teflon-coated magnetic stirring bar and a pressure equalizing dropping funnel topped with a rubber septum, is charged with 0.50 g (1.39 mmol) of $N_3P_3Az_2Cl_4$ and 15 mL of benzene, dried over molecular sieves (4 Å).

A 1 to 1.5 M solution of methylamine in benzene is prepared by leading methylamine (from a lecture bottle) through 100 mL of dry benzene during about 30 min. The actual concentration of the amine is determined by acid–base titration, using dilute hydrochloric acid and bromocresol green as indicator.

A quantity of the amine solution, equivalent to 41.7 mmol of methylamine, is transferred to the dropping funnel by syringe. Under vigorous stirring the amine solution is added dropwise to the contents of the flask, cooled at 6°. The mixture is allowed to warm slowly to room temperature. After the mixture has been stirred for 18 hr, the reaction temperature is

TABLE I ^{31}P $\{^1H\}$ **NMR Data for Some Compounds** $N_3P_3Az_{6-n}$[a]

Compound	δ_{PAz_2}	δ_{PAzCl}	δ_{PCl_2}	$^2J_{PP}$
$N_3P_3AzCl_5$		31.2	22.2	39.0
trans-$2,4\text{-}N_3P_3Az_2Cl_4$		34.7	25.0	38.0
$2,2\text{-}N_3P_3Az_2Cl_4$	34.2		21.9	30.0
cis-$2,4\text{-}N_3P_3Az_2Cl_4$		34.1	24.9	38.2
$2,2,4\text{-}N_3P_3Az_3Cl_3$	35.8	35.8	24.9	AA′X spectrum, multiplet splitting 34.0 Hz

[a]Chemical shifts (ppm) are positive in low field direction, coupling constants are given in hertz (Hz).

raised to 50° during an additional period of 24 hr (using *trans*-2,4- or 2,2-$N_3P_3Az_2Cl_4$ as the starting material) or 48 hr (using *cis*-2,4-$N_3P_3Az_2Cl_4$ as starting material), while stirring is continued. The amine HCl salt formed is removed by filtration and washed three times with 10-mL portions of benzene. The combined benzene fractions are evaporated to dryness *in vacuo*. The crude product thus obtained is recrystallized from a pentane–benzene mixture. As compounds $N_3P_3Az_2(NHMe)_4$ tend to adsorb solvent molecules, the crystals are kept at 45° and low pressure till free from the solvent used. (Control by means of ^1H NMR.)

trans-2,4-$N_3P_3Az_2(NHMe)_4$. mp 153–155°, yield 0.35 g (1.03 mmol = 74%).

Anal. Calcd. for $C_8H_{24}N_9P_3$: C, 28.32; H, 7.13; N, 37.16. Found: C, 28.21; H, 7.18; N, 36.67.

2,2-$N_3P_3Az_2(NHMe)_4$. mp 147–148°, yield 0.30 g (0.88 mmol = 64%).

Anal. Calcd. for $C_8H_{24}N_9P_3$: C, 28.32; H, 7.13; N, 37.16. Found: C, 28.47; H, 7.20; N, 36.59.

cis-2,4-$N_3P_3Az_2(NHMe)_4$. mp 152–153°, yield 0.19 g (0.56 mmol = 40%).

Anal. Calcd. for $C_8H_{24}N_9P_3$: C, 28.32; H, 7.13; N, 37.16. Found: C, 28.27; H, 7.11; N, 36.61.

Properties

Isomers $N_3P_3Az_2(NHMe)_4$ are white, crystalline solids, stable in air at ambient temperature. They are readily soluble in water; in acidic medium decomposition takes place as a consequence of the opening of the aziridinyl rings. ^{31}P {^1H} NMR data (for experimental details see Section A.1.) are given in Table II.

All three isomers exhibit a significant *in vivo* cytostatic activity in the order *trans*-2,4 ≈ *cis*-2,4 > 2,2. The 50% lethal dose (LD_{50}) of *trans*-2,4-

TABLE II ^{31}P {^1H} NMR Data for
Compounds $N_3P_3Az_2(NHMe)_4$

Compound	δ_{Az_2}	$\delta_{PAzNMMe}$	$\delta_{P(NHMe)_2}$	$^2J_{PP}$
trans-2,4-$N_3P_3Az_2(NHMe)_4$		30.0	22.1	37.4
2,2-$N_3P_3Az_2(NHMe)_4$	38.2		22.8	37.9
cis-$N_3P_3Az_2(NHMe)_4$		29.9	22.0	38.5

$N_3P_3Az_2(NHMe)_4$ for mice amounts to 50 mg kg^{-1}. The antitumor effect of this compound on leukemia L 1210 cells in mice expressed as T/C value (T is the mean life span of treated mice; C is the mean life span of control mice) is 200% for a dose equal to 30 mg kg^{-1} with 25% long term survivors.[5]

**B. *trans*-NON-*gem*-BIS(AZIRIDINYL)-
    HEXAKIS(METHYLAMINO)-CYCLOTETRAPHOSPHAZENE
    [*trans*-2,6-BIS(1-AZIRIDINYL)-2,4,4,6,8,8-
    HEXAKIS(METHYLAMINO)-1,3,5,7,2λ^5,4λ^5,6λ^5,8λ^5-
    TETRAZATETRAPHOSPHOCINE]**

$$(NPCl_2)_4 \xrightarrow{\text{(CH}_2)_2\text{NH}} N_4P_4AzCl_7 + \text{isomers } N_4P_4Az_2Cl_6$$
$$+ \text{ isomers } N_4P_4Az_3Cl_5 + (CH_2)_2NH \cdot HCl \quad (3)$$

$$trans\text{-}2,6\text{-}N_4P_4Az_2Cl_6 + 12MeNH_2 \longrightarrow$$
$$trans\text{-}2,6\text{-}N_4P_4Az_2(NHMe)_6 + 6MeNH_2 \cdot HCl \quad (4)$$

$$trans\text{-}2, 6\text{-}N_4P_4Az_2 (NHMe)_6$$

Procedure

■ **Caution.** *See **CAUTION** in Section A. Compounds $N_4P_4Az_{8-n}R_n$ are toxic; direct contact of these solids with the skin must be avoided.*

The reactions should be carried out in a dry nitrogen atmosphere, in order to avoid hydrolysis of partially aminolyzed products.

1. Compounds $N_4P_4Az_{8-n}Cl_n$

A similar procedure as described under Section A.1. follows. To a vigorously stirred solution of 9.3 g (20.1 mmol) of $(NPCl_2)_4$ (ref. 6) in 300 mL of hexane, dried over sodium wire, is added dropwise a solution of 4.2 mL (81.1 mmol) of aziridine in 100 mL of hexane, cooled at 0°. The reaction mixture is warmed slowly to room temperature and stirred for an additional period of 18 hr. The work-up procedure (filtration and washing with hexane) as described under Section A.1. yields a viscous oil. The separation of the reaction products is carried out by HPLC, using a Lichrosorb Si 60/10 column (Chrompack, O.D. 22 mm, length 30 cm) and a 3:1 mixture of hexane–diethyl ether as eluent. Eight fractions are obtained.

Fraction 1. $N_4P_4AzCl_7$, recrystallized from hexane–diethyl ether, mp 68.5–70°, Yield: 1.60 g (3.40 mmol = 16.9%).

Anal. Calcd. for $C_2H_4N_5P_4Cl_7$: C, 5.11; H, 0.86; N, 14.90; Cl, 52.78. Found: C, 5.07; H, 0.84; N, 14.86; Cl, 52.60.

Fraction 2. trans-2,6-$N_4P_4Az_2Cl_6$, recrystallized from hexane–diethyl ether, mp 103–104°, Yield: 0.57 g (1.20 mmol = 6.0%).

Anal. Calcd. for $C_4H_8N_6P_4Cl_6$: C, 10.08; H, 1.69; N, 17.63; Cl, 44.62. Found: C, 10.11; H, 1.60; N, 17.56; Cl, 44.63.

Fraction 3. cis-2,6-$N_4P_4Az_2Cl_6$, recrystallized from hexane–diethyl ether, mp 122.5–123.5°, Yield: 0.38 g (0.80 mmol = 4.0%).

Anal. Calcd. for $C_4H_8N_6P_4Cl_6$: C, 10.08; H, 1.69; N, 17.63; Cl, 44.62. Found: C, 10.08; H, 1.61; N, 17.66; Cl, 44.64.

Fraction 4. trans-2,4-$N_4P_4Az_2Cl_6$, recrystallized from hexane–diethyl ether, mp 91–92°, Yield: 0.76 g (1.59 mmol = 7.9%).

Anal. Calcd. for $C_4H_8N_6P_4Cl_6$: C, 10.08; H, 1.69; N, 17.63; Cl, 44.62. Found: C, 10.21; H, 1.68; N, 17.73; Cl, 44.29.

Fraction 5. 2,2-$N_4P_4Az_2Cl_6$, recrystallized from hexane, mp 39.5–40.5°, Yield: 0.19 g (0.40 mmol = 2.0%).

Anal. Calcd. for $C_4H_8N_6P_4Cl_6$: C, 10.08; H, 1.69; N, 17.63; Cl, 44.62. Found: C, 10.02; H, 1.62; N, 17.57; Cl, 44.94.

Fraction 6. cis-2,4-$N_4P_4Az_2Cl_6$, recrystallized from hexane–diethyl ether, mp 68–70°, Yield: 0.57 g (1.20 mmol = 5.9%).

Anal. Calcd. for $C_4H_8N_6P_4Cl_6$: C, 10.08; H, 1.69; N, 17.63; Cl, 44.62. Found: C, 10.43; H, 1.66; N, 17.47; Cl, 44.53.

Fraction 7. A mixture of 2,2,6-N$_4$P$_4$Az$_3$Cl$_5$ and *cis*-2,6-*trans*-4-N$_4$P$_4$Az$_3$Cl$_5$ (1.08 g).

Fraction 8. *cis*-2,4-*trans*-6-N$_4$P$_4$Az$_3$Cl$_5$, recrystallized from hexane–diethyl ether, mp 84.5–86.5°, Yield: 0.87 g (1.80 mmol = 9.0%).

Anal. Calcd. for C$_6$H$_{12}$N$_7$P$_4$Cl$_5$: C, 14.91; H, 2.50; N, 20.28; Cl, 36.67. Found: C, 14.72; H, 2.57; N, 20.41; Cl, 36.96.

Properties

Compounds N$_4$P$_4$AzCl$_7$, N$_4$P$_4$Az$_2$Cl$_6$, and *cis*-2,4-*trans*-6-N$_4$P$_4$Az$_3$Cl$_5$ are white, crystalline solids, which are rather sensitive to hydrolysis and should be kept in a dry atmosphere at low temperature. ^{31}P {^{1}H} NMR data (for experimental details see Section A.1.) are given in Table III.

The crystal structure of *cis*-2,4-N$_4$P$_4$Az$_2$Cl$_6$ has been determined.[10]

2. Compound *trans*-2,6-N$_4$P$_4$Az$_2$(NHMe)$_6$

A similar procedure as described under Section A.2. follows. To a vigorously stirred solution of 0.48 g (1.0 mmol) of *trans*-2,6-N$_4$P$_4$Az$_2$Cl$_6$ in 15 mL of chloroform (dried over 4 Å molecular sieves), cooled at 0°, is added dropwise a 20-fold excess of methylamine using about a 1 *M* solution in dry benzene. The reaction mixture is warmed slowly to room temperature and stirred for an additional period of 18 hr at that temperature. The usual work-up procedure (Section A.2.) affords a white solid, which is recrystallized several times from a mixture of diethyl ether and dichloromethane. Yield: 0.25 g (0.56 mmol = 56%), mp 124–126°.

Anal. Calcd. for C$_{10}$H$_{32}$N$_{12}$P$_4$: C, 27.03; H, 7.26; N, 37.83. Found: C, 26.90; H, 7.26; N, 37.43.

TABLE III ^{31}P {^{1}H} **NMR Data for Some Compounds N$_4$P$_4$Az$_{8-n}$Cl$_n$**

Compound	$\delta_{P(2)}$	$\delta_{P(4)}$	$\delta_{P(6)}$	$\delta_{P(8)}$	J_{24}	J_{46}	J_{68}	J_{28}
N$_4$P$_4$AzCl$_7$	8.6	−4.7	−7.2	−4.7	27.6	30.6	30.6	27.6
trans-2,6-N$_4$P$_4$Az$_2$Cl$_6$	8.4	−1.9	8.4	−1.9	27.9	27.9	27.9	27.9
cis-2,6-N$_4$P$_4$Az$_2$Cl$_6$	8.7	−2.6	8.7	−2.6	28.4	28.4	28.4	28.4
trans-2,4-N$_4$P$_4$Az$_2$Cl$_6$	11.8	11.8	−4.9	−4.9	27.6	25.4	31.1	25.4
2,2-N$_4$P$_4$Az$_2$Cl$_6$	18.8	−5.9	−6.5	−5.9	11.6	26.1	26.1	11.6
cis-2,4-N$_4$P$_4$Az$_2$Cl$_6$	10.3	10.3	−5.0	−5.0	29.2	27.1	32.7	27.1
cis-2,4-*trans*-6-N$_4$P$_4$Az$_3$Cl$_5$	10.3	13.7	11.7	−1.8	28.9	27.6	24.7	26.9

Properties

The compound *trans*-2,6-$N_4P_4Az_2$(NHMe)$_6$ is a white solid, stable in air at ambient temperature. It is readily soluble in water; in acidic medium a rapid decomposition occurs by opening of the aziridinyl rings. The ^{31}P {^1H} NMR data (80.9 MHz, CDCl$_3$) are $\delta_{PAzNHMe} = 13.6$, $\delta_{P(NHMe)_2} = 9.5$ ppm, $^2J_{PP} = 32.3$ Hz.

The 50% lethal dose (LD$_{50}$) for mice amounts to 225 mg kg^{-1}. The antitumor effect on L 1210 in mice, expressed as T/C value is 244% for a dose of 100 mg kg^{-1} with 40% long term survivors.[5]

References

1. A. A. van der Huizen, J. C. van de Grampel, W. Akkerman, P. Lelieveld, A. van der Meer-Kalverkamp, and H. B. Lamberts, *Inorg. Chim. Acta*, **78**, 239 (1983).
2. J. C. van de Grampel, A. A. van der Huizen, A. P. Jekel, J. W. Rusch, T. Wilting, W. Akkerman, P. Lelieveld, H. B. Lamberts, A. van der Meer-Kalverkamp, N. H. Mulder, and S. Rodenhuis, *Phosphorus Sulfur,* **18**, 337 (1983).
3. A. A. van der Huizen, Ph.D. Thesis, University of Groningen, Groningen, The Netherlands, 1984.
4. Netherlands Patent Application 8300573 (1983); Netherlands Patent Application 8301626 (1983).
5. A. A. van der Huizen, T. Wilting, J. C. van de Grampel, P. Lelieveld, A. van der Meer-Kalverkamp, H. B. Lamberts, and N. H. Mulder, *J. Med. Chem.,* **29**, 1341 (1986).
6. M. L. Nielsen and G. Cranford, *Inorg. Synth.,* **6**, 94 (1960).
7. H. Wenker, *J. Am. Chem. Soc.,* **57**, 2328 (1935).
8. P. A. Leighton, W. A. Perkins, and M. L. Renquist, *J. Am. Chem. Soc.,* **69**, 1540 (1974).
9. W. A. Reeves, G. L. Drake, Jr., and C. L. Hoffpauir, *J. Am. Chem. Soc.,* **73**, 3522 (1951).
10. A. A. van der Huizen, J. C. van de Grampel, J. W. Rusch, T. Wilting, F. van Bolhuis, and A. Meetsma, *J. Chem. Soc. Dalton Trans.* **1986**, 1317.

20. *cis*-DIAMMINEPLATINUM α-PYRIDONE BLUE

Submitted by PRADIP K. MASCHARAK* and STEPHEN J. LIPPARD†
Checked by F. ALBERT COTTON‡ and DANIEL P. BANCROFT‡

Although the isolation of the first "blue" platinum compound was reported as early as 1908,[1] the nature of the bonding in the "platinum blues" re-

*Department of Chemistry, University of California, Santa Cruz, CA 95064.
†Department of Chemistry, Massachusetts Institute of Technology, Cambridge, MA 02139. Address correspondence to this author.
‡Department of Chemistry, Texas A&M University, College Station, TX 77843-3255.

mained unknown for many decades. Despite various physical measurements, which indicated polymeric structures with bridging amidate type of linkages for these paramagnetic blue species, definitive structural information was elusive. The report that blue platinum complexes with pyrimidines and substituted amides exhibit a high index of antitumor activity with low nephrotoxicity[2] further stimulated interest in the structure and chemistry of these intriguing blue species. The major advance came in 1977 when *cis*-diammine platinum α-pyridone blue was crystallized and the structure was solved by X-ray diffraction.[3] Subsequently, two more tetranuclear crystalline blues have been characterized, one of which contains 1-methyluracil ligands.[4] All three crystalline derivatives are comprised of amidate-bridged, tetrameric platinum chains with partial metal–metal bonding and mixed oxidation states. The α-pyridone blue has been investigated most thoroughly[5] and studies in aqueous solution have led to the isolation of related dimeric and monomeric complexes.[6] In aqueous solution, one equivalent of reducing agent splits the tetrameric chain into two platinum(II) dimers where the two metals are held by α-pyridone in a head-to-head fashion. On the other hand, oxidation with 3 to 5 M nitric acid produces a metal–metal bonded stable diamagnetic Pt(III) dimer. Both of these dimers have been structurally characterized.[6,7] In spite of their detailed physical characterization, however, reproducible, high yield synthetic procedures for α-pyridone blues were not available. Two synthetic procedures are reported in detail in this paper.

A. BIS[BIS(μ-2-PYRIDONATO-*N*1,*O*2)BIS(*cis*-DIAMMINEPLATINUM(2.25+))] PENTANITRATE HYDRATE (*cis*-DIAMMINEPLATINUM α-PYRIDONE BLUE OR PPB)

Method A

$$cis\text{-}[(NH_3)_2Pt(H_2O)_2]^{2+} + C_5H_5NO \xrightarrow[\text{room T}]{\text{water, pH 5.5}} \text{greenish blue solution}$$

$$\xrightarrow[\text{NaNO}_3]{7\ M\ \text{HNO}_3\ \text{pH}\leq1} [Pt_2(NH_3)_4(C_5H_4NO)_2]_2(NO_3)_5 \cdot H_2O$$

Procedure

2 g (4 mmol) of *cis*-[(NH$_3$)$_2$PtI$_2$] (ref. 8) is allowed to react with 1.35 g (8 mmol) of silver nitrate in 35 mL of water at 60° for 1 hr and the mixture is filtered through Celite® to remove silver iodide. A 400 mg (4.2 mmol)

sample of α-pyridone, recrystallized from benzene, is then added to the pale yellow filtrate (pH ~ 3) containing cis-$[(NH_3)_2Pt(H_2O)_2]^{2+}$ and the pH is adjusted to 5.5 with 1 M sodium hydroxide solution. The clear yellow solution is stirred in a stoppered flask at room temperature for 20 hr. During this period the color changes to light green. The pH is lowered to 1 with 7 M nitric acid and 1 g of sodium nitrate is added. Cooling the deep blue solution to 0° results in dark shiny needles of PPB. The crystals are dichroic and, when immersed in solution, appear magenta under reflected light. They are filtered, washed with small portions (4–5 mL) of cold 0.1 M nitric acid solution, and dried over $CaSO_4$. Yield: 450 mg (27%). Checker's yield (10%).

Method B

$$cis\text{-}[(NH_3)_2Pt(C_5H_5NO)_2]^{2+} + cis\text{-}[(NH_3)_2Pt(H_2O)_2]^{2+} \xrightarrow[\text{room T}]{\text{water pH 5.5}}$$

$$\begin{matrix} \text{greenish} \\ \text{blue} \\ \text{solution} \end{matrix} \xrightarrow[\text{NaNO}_3]{7\ M\ \text{HNO}_3,\ \text{pH}\leq1} [Pt_2(NH_3)_4(C_5H_4NO)_2]_2(NO_3)_5{\cdot}H_2O$$

Procedure

The filtrate from the reaction of 2 g (4 mmol) of cis-$[(NH_3)_2PtI_2]$ and 1.35 g (8 mmol) of silver nitrate in 35 mL of water at 60° (as above) is allowed to react with 800 mg (8.5 mmol) of α-pyridone at 50° under nitrogen for 24 hr. The pale green reaction mixture is then evaporated to dryness at 50° on a rotary evaporator leaving pale blue needles of $[(NH_3)_2Pt(C_5H_5NO)_2](NO_3)_2$.[6] To this residue is added the filtrate from another reaction of 2 g of cis-$[(NH_3)_2PtI_2]$ and 1.35 g of silver nitrate in 35 mL of warm water. The pH of the resulting mixture is adjusted to 5.5 with 1 M sodium hydroxide solution and stirred at room temperature for 24 hr. The small amount (~80 mg) of white residue which appears during this period is filtered off and the pH is then adjusted to 1 with 7 M nitric acid. The deep blue solution is cooled at 0° for 20 hr. The first batch of PPB crystals (400 mg) is collected and the filtrate is kept again at 0° after addition of 2 g of sodium nitrate. Crystals (650 mg) of PPB are collected after 24 hr. The two batches are combined and washed with cold 0.1 M nitric acid solution and dried over $CaSO_4$. Total yield 1.05 g (32%) (checker's yield 31%).

The product obtained from these procedures is pure. However, PPB can be recrystallized from 0.1 M nitric acid solution to produce dark plate-like crystals.

Properties

Crystals of PPB are stable in air and dissolve in water to give a deep blue solution, which slowly bleaches with time. The blue color is enhanced and stabilized in the presence of nitric or perchloric acid (pH ~ 1) and/or 0.1 M sodium nitrate or perchlorate. In 0.1 M nitric acid PPB exhibits three optical spectroscopic bands at 680, 480, and 405 nm. The blue color is rapidly discharged in the presence of chloride ion and can be restored by removing chloride with silver nitrate. *cis*-Diammineplatinum α-pyridone blue is paramagnetic and exhibits strong ESR signals at room temperature, with g values of 2.37 and 1.99.

References

1. K. A. Hofmann and G. Bugge, *Chem. Ber.*, **41**, 312 (1908).
2. P. J. Davidson, P. J. Faber, R. G. Fischer, Jr., S. Mansy, H. J. Peresie, B. Rosenberg, and L. van Camp, *Cancer Chemother. Rep.*, **59**, 287 (1975); R. J. Speer, H. Ridgeway, L. M. Hall, D. P. Stewart, K. E. Howe, D. Z. Lieberman, A. D. Newman, and J. M. Hill, *Cancer Chemother. Rep.*, **59**, 629 (1975).
3. J. K. Barton, H. N. Rabinowitz, D. J. Szalda, and S. J. Lippard, *J. Am. Chem. Soc.*, **99**, 2827 (1977).
4. T. V. O'Halloran, M. M. Roberts, and S. J. Lippard, *J. Am. Chem. Soc.*, **106**, 6427 (1984); P. K. Mascharak, I. D. Williams, and S. J. Lippard, *J. Am. Chem. Soc.*, **106**, 6428 (1984); T. V. O'Halloran, P. K. Mascharak, I. D. Williams, M. M. Roberts, and S. J. Lippard, *Inorg. Chem.*, **26**, 1261 (1987).
5. J. K. Barton, S. A. Best, S. J. Lippard, and R. A. Walton, *J. Am. Chem. Soc.*, **100**, 3785 (1978); J. K. Barton, D. J. Szalda, H. N. Rabinowitz, J. V. Waszczak, and S. J. Lippard, *J. Am. Chem. Soc.*, **101**, 1434 (1979); J. K. Barton, C. Caravana, and S. J. Lippard, *J. Am. Chem. Soc.*, **101**, 7269 (1979).
6. L. S. Hollis and S. J. Lippard, *J. Am. Chem. Soc.*, **103**, 1230 (1981); **105**, 3494 (1983); *Inorg. Chem.*, **22**, 2708 (1983).
7. L. S. Hollis and S. J. Lippard, *J. Am. Chem. Soc.*, **103**, 6761 (1981); *Inorg. Chem.*, **21**, 2116 (1982).
8. S. G. Dhara, *Indian J. Chem.*, **8**, 193 (1970).

Chapter Four

METAL COMPOUNDS, COMPLEXES, AND LIGANDS

21. DIPOTASSIUM TETRAIODOPLATINATE(II) DIHYDRATE

$$K_2[PtCl_4] + 4KI + 2H_2O \longrightarrow K_2[PtI_4]\cdot 2H_2O + 4KCl$$

Submitted by LARS-FRIDE OLSSON*
Checked by C. KING† and D. MAX ROUNDHILL†

The potassium tetrachloro[1,2] and tetrabromo complexes[2,3] of platinum(II) have been used as starting materials for a large number of platinum(II) complexes. The iodo complex, $K_2[PtI_4]$ is difficult to synthesize[4] for three reasons: the extremely low solubility of platinum(II) iodide, PtI_2, the ease of formation[5] of $[Pt_2I_6]^{2-}$ ($2[PtI_4]^{2-} \rightarrow [Pt_2I_6]^{2-} + 2I^-$), and the low resistance towards oxidation to platinum(IV) by triiodide and iodine. The latter two are formed by air oxidation of iodide. These problems can be circumvented by keeping iodide in large excess over platinum, by continuously working under a nitrogen atmosphere, and by preparing all aqueous

*Inorganic Chemistry 1, Chemical Center, University of Lund, P.O. Box 124, S-221 00 Lund, Sweden.
†Department of Chemistry, Tulane University, New Orleans, LA 70118.

solutions from deoxygenated water. The title compound is separated from the excess KI by slow crystallization.

Procedure

Into a 125-mL Erlenmeyer flask, which is continuously flushed with nitrogen, place a magnetic stirring bar, 1.0 g (2.4 mmol) of finely divided $K_2[PtCl_4]$ and 5.0 mL of deoxygenated water. Prepare about 10 mL of a saturated aqueous solution of KI (flush with nitrogen). When all $K_2[PtCl_4]$ has dissolved, add 4 mL of KI (aq, sat). The solution becomes dark and a small amount of material precipitates. The flask is stored under nitrogen for 15 hr.

Upon addition of 50 mL of acetone, with stirring, KCl is precipitated quantitatively. The solution is decanted to a 250 mL round-bottom flask. The precipitate is washed with 4 × 25 mL portions of acetone, which are decanted to the flask. The last washing solution should be slightly yellow and the precipitate white. The solution is evaporated on a rotary evaporator with the aid of a water bath kept at 25°, until the volume is 5 to 15 mL.

The solution is transferred with a disposable pipet to a 200-mL conical beaker, aided by a few drops of water. The beaker is placed in a nitrogen filled desiccator. The desiccator lid is fitted with two glass tubes in a two-hole stopper. The nitrogen inlet tube extends to the bottom of the desiccator. The outlet tube is flush with the bottom of the stopper. After purging the desiccator with nitrogen for 15 min, the flow is reduced to one bubble every 2 sec from an oil bubbler. The desiccator is left at room temperature for 1 to 4 weeks.

When the flask contents are dry, excess KI, together with some $K_2[PtI_4]\cdot 2H_2O$ has crept up the wall of the beaker. A few large, thin, black, rhombic crystals of the title compound are formed on the bottom.

It may be necessary to mechanically remove solid KI from the crystal surface. The yield is 1.2–1.4 g (60–71%).

Anal. Calcd. for $H_4I_4K_2O_2Pt$: I, 62.1; K, 9.6; Pt, 23.9%. Found: I, 62.1; K, 9.1; Pt, 23.5%.

Properties

The compound (MW = 816.9 g mol^{-1}) is moderately stable and can be handled in air for some hours. It should be stored under a dry, inert atmosphere and protected from light. It is soluble in alcohol, acetone, and acetonitrile, but at room temperature it undergoes rapid conversion[5] to $[Pt_2I_6]^{2-}$. In aqueous solution it is slowly precipitated as PtI_2. The electronic spectrum in aqueous solution[4,5] containing a large excess of iodide is characterized by λ_{max} 387 nm, ϵ 5.04 $\times$ 10^3 M^{-1} cm; λ_{max} 331, ϵ 8.3 $\times$ 10^3 and λ_{min} 299, ϵ 4.5 $\times$ 10^2. The platinum -195 NMR spectrum in D_2O with added KI shows a single resonance at $\delta = -5448$ ppm (H_2PtCl_6 reference, $\delta = 0$) and with a relaxation time $T_1 = 0.55(2)$ sec at 4.7 tesla.[6]

Note: It is advisable to check for PtI_6^{2-}, either by electronic absorption or by the platinum-195 NMR spectrum. The molar absorbtivity at 495 nm, where PtI_6^{2-} has a maximum with ϵ 1.1 $\times$ 10^4 M^{-1} cm^{-1}, should not exceed 200 M^{-1} cm^{-1}. The absence of PtI_6^{2-} can also be verified by the absence of its peak at $\delta = -6291$ ppm.[6]

References

1. G. B. Kauffman and D. O. Cowan, *Inorg. Synth.*, **7**, 240 (1963).
2. S. E. Livingstone, *Synth. Inorg. Metal Org. Chem.*, **1**, 1 (1971).
3. J. A. Abys, N. P. Enright, H. M. Gerdes, T. L. Hall, and J. M. Williams, *Inorg. Synth.*, **19**, 2 (1979).
4. B. Corain and A. J. Poë, *J. Chem. Soc.*, **1967**(A), 1318.
5. L. F. Olsson, *Inorg. Chem.*, **25**, 1697 (1986).
6. The NMR data are reported by the checkers.

22. ALKYL OR ARYL BIS(TERTIARY PHOSPHINE) HYDROXO COMPLEXES OF PLATINUM(II)

Submitted by M. A. BENNETT* and A. ROKICKI*
Checked by R. J. PUDDEPHATT†

Complexes of the type $[Pt(OH)RL_2]$ (R = alkyl or aryl; L_2 = two monodentate tertiary phosphines or a bidentate ditertiary phosphine) are rare examples of monomeric, uncharged hydroxo complexes of platinum(II). Their interesting chemical properties arise from the presence of typical

*Research School of Chemistry, The Australian National University, Canberra, A.C.T., Australia 2601.

†Department of Chemistry, University of Western Ontario, London, Ontario, Canada N6A 5B7.

covalently binding ligands together with the electronegative ligand OH in the coordination sphere of a soft or class B metal ion. The ready substitution of the hydroxo ligand in the presence of a variety of strong and weak acids, both organic and inorganic, provides a useful synthesis of a range of platinum(II) complexes, including those containing functionalized alkyl groups.[1-7] The Pt—OH bond also undergoes insertion with many simple molecules (e.g., CO, isocyanides, CO_2, COS, CS_2, and SO_2),[6,8,9] and its reaction with nitriles to form N-amidoplatinum(II) complexes enables hydroxoplatinum(II) complexes to catalyze homogeneously the hydration of nitriles to carboxamides.[3,7,10]

Some hydroxoplatinum(II) complexes have been made by hydrolysis of the corresponding chloroplatinum(II) compounds [PtClRL$_2$] with alkali metal hydroxides,[2] and the presence of a crown ether may be beneficial.[11] The most general procedure, however, is first to abstract chloride ion from the precursor with Ag[BF$_4$] in a suitable solvent and then to treat the resulting solvent species [PtR(solvent)L$_2$] [BF$_4$] with aqueous NaOH or KOH. The choice of solvent is limited by the reactivity of the product hydroxo complex. Thus, although acetone is generally a convenient solvent for chloride abstraction, it cannot be used in the second step for making [Pt(OH)(CH$_3$)(dppp)]* because this hydroxo complex reacts rapidly with acetone to give [Pt(CH$_2$COCH$_3$)(CH$_3$)(dppp)].[7] Likewise, methanol cannot be used in the synthesis of *trans*-[Pt(OH)RL$_2$] complexes because of rapid reaction to give *trans*-[PtH(R)L$_2$], among other products.[12]

Two general precautions should be observed in the four representative syntheses described here. First, the chloride abstraction step is best carried out with exclusion of light to prevent decomposition of the precipitated AgCl; this is done by wrapping the flask with aluminum foil. Second, the temperature of reaction mixtures should not be allowed to exceed ~35°, especially during the removal of solvents under reduced pressure, otherwise poorly soluble solids of unknown composition are formed and the yields of desired hydroxo complexes are greatly reduced. All the reactions have been carried out under nitrogen with use of standard Schlenk techniques and degassed solvents. Of the compounds described, however, only *trans*-[Pt(OH)(C$_6$H$_5$)(PEt$_3$)$_2$] is appreciably air sensitive, as a result of its rapid absorption of CO_2 to form a bicarbonato complex. The precursor complexes [PtClRL$_2$] are well-established compounds that are most conveniently synthesized by displacement of 1,5-cyclooctadiene from [PtClR(cod).][7,13]

---

*Abbreviations: dppp = 1,3-propanediylbis(diphenylphosphine), Ph$_2$P[CH$_2$]$_3$PPh$_2$; cy = cyclohexyl, C$_6$H$_{11}$; cod = cyclooctadiene; THF = tetrahydrofuran, solvent; thf = tetrahydrofuran, ligand.

A.   *trans*-HYDROXO(PHENYL)BIS(TRIETHYL-PHOSPHINE)PLATINUM(II)

trans-[PtCl(CH$_3$)(Pcy$_3$)$_2$] + Ag[BF$_4$] $\xrightarrow{\text{THF/acetone}}$

$$\text{\textit{trans}-[Pt(CH}_3\text{)(solvent)(Pcy}_3\text{)}_2\text{][BF}_4\text{)} + \text{AgCl}$$

trans-[Pt(CH$_3$)(solvent)(Pcy$_3$)$_2$][BF$_4$] + KOH $\longrightarrow$

$$\text{\textit{trans}-[Pt(OH)(CH}_3\text{)(Pcy}_3\text{)}_2\text{]} + \text{K[BF}_4\text{]}.$$

Procedure

A mixture of [PtCl(C$_6$H$_5$)(PEt$_3$)$_2$] (1.36 g, 2.5 mmol) and Ag[BF$_4$](0.49 g, 2.5 mmol) is placed in a 100-mL Schlenk flask or round-bottom flask fitted with a nitrogen inlet and a magnetic stirring bar and THF (50 mL) is added. The mixture is stirred for 1 hr at room temperature and the precipitated AgCl is removed by centrifugation. The resulting colorless solution is stirred vigorously for 0.5 hr with a solution of KOH(0.16 g, 2.85 mmol) in water (10 mL). The solution is again centrifuged to remove a small amount of gray solid, and the supernatant liquid is evaporated *in vacuo* to give a pale olive oil, which is extracted with diethyl ether (3 × 10 mL). The extracts are evaporated *in vacuo* to yield a colorless, viscous oil that is crystallized from isopentane (~10 mL) at −78°. The mother liquor is decanted and the product is dried *in vacuo* at room temperature, yielding a white solid (1.05 g, 2.0 mmol, 80%).

Anal. Calcd. for C$_{18}$H$_{36}$OP$_2$Pt: C, 41.1; H, 6.9; P, 11.8. Found: C, 40.8; H, 6.6; P, 10.7.

Properties

Prepared as just described, *trans*-[Pt(OH)(C$_6$H$_5$)(PEt$_3$)$_2$] remains solid under dry nitrogen, but depending on the temperature of the surroundings it readily softens and melts at about 30°. The complex liquefies rapidly in moist air and is very soluble in most organic solvents. The IR spectrum shows a weak ν_{OH} band at ~3670 cm^{-1} in CH$_2$Cl$_2$, but this cannot be seen in Nujol mull or KBr disk spectra. The OH resonance cannot be located in the ^{1}H NMR spectrum, which does not differ much from that of *trans*-[PtCl(C$_6$H$_5$)(PEt$_3$)$_2$]. In C$_6$D$_6$ there is a multiplet in the region δ = 7.50–7.01 ppm due to the phenyl protons and overlapping multiplets at δ = 1.51 and 0.97 ppm due to the ethyl protons; the last two appear at δ = 1.46 and 0.96 ppm in toluene-d_8. The ^{31}P {^{1}H} NMR spectrum of an approximately 0.03 M solution in CD_2Cl_2 at 21°, measured at 80.98 MHz on a Bruker CXP200 instrument, shows a singlet at δ = 16.0 ppm (to high frequency of external 85% H_3PO_4) with ^{195}Pt satellites ($^1J_{PtP}$ = 2884 Hz); in toluene-d_8 these values are δ = 15.9 ppm ($^1J_{PtP}$ = 2912 Hz). Both the chemical shift and the coupling constant may vary somewhat with solvent, concentration, and temperature.

B. *trans*-HYDROXO(PHENYL)BIS(TRIPHENYL-PHOSPHINE)PLATINUM(II)

trans-$[PtCl(C_6H_5)(PPh_3)_2]$ + $Ag[BF_4]$ $\xrightarrow{\text{acetone}}$

$\qquad$ *trans*-$[Pt(C_6H_5)(acetone)(PPh_3)_2][BF_4]$ + AgCl

trans-$[Pt(C_6H_5)(acetone)(PPh_3)_2][BF_4]$ + KOH $\longrightarrow$

$\qquad$ *trans*-$[Pt(OH)(C_6H_5)(PPh_3)_2]$ + $K[BF_4]$

Procedure

A 50-mL Schlenk flask fitted with a magnetic stirring bar is charged with *trans*-$[PtCl(C_6H_5)(PPh_3)_2]$ (0.16 g, 0.20 mmol) and acetone (15 mL). The suspension is treated with a solution of $Ag[BF_4]$(0.039 g, 0.20 mmol) in acetone (5 mL) and the mixture is stirred at room temperature for 3 hr. The precipitated AgCl is removed by centrifugation and washed with acetone (10 mL). The combined acetone solutions are stirred with a solution of KOH (0.014 g, 0.24 mmol) in water (1 mL) for 2 hr at room temperature, during which time the color turns pale yellow and a small amount of colorless precipitate is formed. The mixture is evaporated to dryness under reduced pressure and the pale yellow solid residue is extracted with benzene (3 × 10 mL). Concentration of the benzene extract and addition of hexane (~10 mL) gives the product as a white, microcrystalline solid (0.12 g, 0.148 mmol, 74%).

Anal. Calcd. for $C_{42}H_{36}OP_2Pt$: C, 62.0; H, 4.5. Found: C, 62.0; H, 4.5.

Properties

The complex *trans*-$[Pt(OH)(C_6H_5)(PPh_3)_2]$ is an air-stable, colorless solid that is readily soluble in CH_2Cl_2, $CHCl_3$, THF, and benzene. No band due to ν_{OH} could be identified in the IR spectrum in a KBr disk, although Otsuka et al.[2] report that the ν_{OH} band appears at 3613 cm^{-1} (Nujol) and at 3619 cm^{-1} (CH_2Cl_2). The ^1H NMR spectrum in CD_2Cl_2 shows the OH resonance as a triplet at δ = 2.16 ppm ($^3J_{PH}$ = 2.5 Hz) flanked by ^{195}Pt satellites ($^2J_{PtH}$ = 19.5 Hz) and the ^{31}P {^{1}H} NMR spectrum in CD_2Cl_2, measured on a Jeol FX60 instrument at 24.29 Hz, shows a singlet at δ = 11.4 ppm (to high frequency of external 85% H_3PO_4) with ^{195}Pt satellites ($^1J_{PtP}$ = 3252 Hz).

C. *trans*-HYDROXO(METHYL)BIS(TRICYCLOHEXYL-PHOSPHINE)PLATINUM(II)

trans-[PtCl(CH$_3$)(Pcy$_3$)$_2$] + Ag[BF$_4$] $\xrightarrow{\text{THF/acetone}}$
$$\text{\textit{trans}-[Pt(CH}_3\text{)(solvent)(Pcy}_3\text{)}_2\text{][BF}_4\text{]} + \text{AgCl}$$

trans-[Pt(CH$_3$)(solvent)(Pcy$_3$)$_2$][BF$_4$] + KOH $\longrightarrow$
$$\text{\textit{trans}-[Pt(OH)(CH}_3\text{)(Pcy}_3\text{)}_2\text{]} + \text{K[BF}_4\text{]}.$$

Procedure

This is similar to that described in the literature for *trans*-[Pt(OH)(C$_6$H$_5$)(Pcy$_3$)$_2$].[7]

A suspension of *trans*-[PtCl(CH$_3$)(Pcy$_3$)$_2$] (0.242 g, 0.30 mmol) in THF (15 mL) is placed in a 50-mL Schlenk flask and treated with Ag[BF$_4$] (0.058 g, 0.30 mmol) dissolved in acetone (2 mL). The mixture is stirred for 0.5 hr. The precipitated AgCl is removed by centrifugation and washed with acetone (5 mL). The combined solutions are stirred at room temperature for 0.5 hr with a solution of KOH (0.019 g, 0.36 mmol) in water (1.5 mL), during which time a small amount of colorless solid is formed. The solution is evaporated to dryness under reduced pressure and the residue is extracted successively with a 20-mL and a 10-mL portion of benzene. The extract is evaporated to ~5-mL volume and hexane is added to give the product as a white, crystalline solid (0.15 g, 0.19 mmol, 63%).

Anal. Calcd. for C$_{37}$H$_{70}$OP$_2$Pt: C, 56.4; H, 8.95; P, 7.9. Found: C, 56.4; H, 9.1; P, 7.2.

Properties

The complex [Pt(OH)(CH$_3$)(Pcy$_3$)$_2$] is an air-stable, colorless, microcrystalline solid that is soluble in CH_2Cl_2 and benzene. We have been unable to locate the ν_{OH} band in the IR spectrum, either in CH_2Cl_2 solution or in a KBr disk. The ^{31}P {^{1}H} NMR spectrum in benzene-d_6, measured at

80.98 MHz on a Bruker CXP200 instrument, shows a singlet at $\delta = 23.9$ ppm (to high frequency of external 85% H_3PO_4) with ^{195}Pt satellites ($^1J_{PtP} = 2927$ Hz).

D. *trans*-HYDROXO(METHYL)[1,3-PROPANEDIYLBIS (DIPHENYLPHOSPHINE)]PLATINUM(II)

$[PtCl(CH_3)(dppp)] + Ag[BF_4] \xrightarrow{\text{THF}}$

$$[Pt(CH_3)(thf)(dppp)][BF_4] + AgCl$$

$[Pt(CH_3)(THF)(dppp)][BF_4] + KOH \longrightarrow$

$$[Pt(OH)(CH_3)(dppp)] + K[BF_4]$$

Procedure

This is a modified version of that given in the literature.[7]

A solution of $[PtCl(CH_3)(dppp)]$(0.33 g, 0.50 mmol) in THF (30 mL) is added to solid $Ag[BF_4]$ (0.11 g, 0.55 mmol) in a 50-mL Schlenk flask; a white precipitate is formed immediately. The mixture is stirred at room temperature for 2 hr and the precipitated AgCl is removed by centrifugation. The precipitate is washed with chloroform (10 mL) and the chloroform solution immediately taken to dryness under reduced pressure, yielding a glassy white solid (the chloroform solution should not be allowed to stand overnight, because the hydroxo complex reacts with $CHCl_3$ to reform $[PtCl(CH_3)(dppp)]$). The THF solution is added to the glassy solid and to the resulting suspension is added with stirring a solution of KOH(0.056 g, 1.00 mmol) in water (2 mL). This immediately gives a colorless solution but within a few minutes a gray solid precipitates. The mixture is stirred for 2 hr, the solid is removed by centrifugation, and the solution is evaporated to dryness under reduced pressure to give a gray solid. Most of this dissolves when it is extracted with benzene (3 × 20 mL). The filtered extract is concentrated to ~10 mL under reduced pressure and hexane is added to give the product as a white solid (0.27 g, 0.42 mmol, 84%).

Anal. Calcd. for $C_{28}H_{30}OP_2Pt$: C, 52.6; H, 4.7; Found: C, 52.6; H, 4.7.

Properties

The complex $[Pt(OH)(CH_3)(dppp)]$ is an air-stable, colorless solid that is very soluble in CH_2Cl_2, $CHCl_3$, and THF, fairly soluble in benzene, and

TABLE I ^{31}P $\{^1H\}$ **NMR Data for**
[Pt(OH)(CH$_3$)(dppp)]^{a,b}

Solvent	$\delta_{(P^1)}(^1J_{PtP^1})$	$\delta_{(P^2)}(^1J_{PtP^2})$	$^2J_{PP}$
C_6D_6	$-0.86(3324)$	$-3.6(1727)$	~21
$CDCl_3$	$-0.91(3511)$	$+3.4(1636)$	19.6
CD_2Cl_2	$-1.1(3403)$	$+2.1(1688)$	~21

^{a}Chemical shifts are relative to external 85% H_3PO_4, positive
to high frequency; coupling constants are in hertz.
^{b}P^1 trans to OH, P^2 trans to CH$_3$.

insoluble in alkanes or diethyl ether. Its ^1H NMR spectrum in CD_2Cl_2
exhibits a doublet of doublets due to PtCH_3 at $\delta = 0.16$ ppm $[^3J_{PH}(cis) =$
3.9 Hz, $^3J_{PH}(trans) = 7.3$ Hz] flanked by ^{195}Pt satellites $[^2J_{PtH} = 60.8$ Hz],
but the signal due to the hydroxyl proton cannot be located. The IR spectrum in CH_2Cl_2 shows a sharp ν_{OH} band at 3601 cm^{-1}, and this appears in
a KBr disk as a weak band at ~3650 cm^{-1} on the high frequency side of
the band due to water in the disk. The ^{31}P $\{^1H\}$ NMR spectrum of
[Pt(OH)(CH$_3$)(dppp)], measured on a Bruker CXP200 instrument at
80.98 MHz, shows a doublet of doublets with ^{195}Pt satellites; the chemical
shifts and coupling constants are given in Table I.

References

1. M. A. Bennett, G. B. Robertson, P. O. Whimp, and T. Yoshida, *J. Am. Chem. Soc.,* **95**, 3028 (1973).
2. T. Yoshida, T. Okano, and S. Otsuka, *J. Chem. Soc., Dalton Trans.,* **1976**, 993.
3. M. A. Bennett and T. Yoshida, *J. Am. Chem. Soc.,* **100**, 1750 (1978).
4. T. G. Appleton and M. A. Bennett, *Inorg. Chem.,* **17**, 738 (1978).
5. R. Ros, R. A. Michelin. R. Bataillard, and R. Roulet, *J. Organomet. Chem.,* **161**, 75 (1978).
6. R. A. Michelin, M. Napoli, and R. Ros, *J. Organomet. Chem.,* **175**, 239 (1979).
7. D. P. Arnold and M. A. Bennett, *J. Organomet. Chem.,* **199**, 119 (1980).
8. R. A. Michelin and R. Ros, *J. Organomet. Chem.,* **169**, C42 (1979).
9. M. A. Bennett and A. Rokicki, *Organometallics,* **4**, 180 (1985).
10. M. A. Bennett and T. Yoshida, *J. Am. Chem. Soc.,* **95**, 3030 (1973).
11. M. E. Fakley and A. Pidcock, *J. Chem. Soc., Dalton Trans.,* **1977**, 1444.
12. D. P. Arnold and M. A. Bennett, *Inorg. Chem.,* **23**, 2110 (1984).
13. H. C. Clark and L. E. Manzer, *J. Organomet. Chem.,* **59**, 411 (1973); M. A. Bennett, R. Bramley, and I. B. Tomkins, *J. Chem. Soc., Dalton Trans.,* **1973**, 166; H. C. Clark and C. R. Jablonski, *Inorg. Chem.,* **14**, 1518 (1975); T. G. Appleton, M. A. Bennett, and I. B. Tomkins, *J. Chem. Soc., Dalton Trans.,* **1976**, 439; C. Eaborn, K. J. Odell, and A. Pidcock, *J. Chem. Soc., Dalton Trans.,* **1978**, 357.

23. TRIS(BIDENTATE)RUTHENIUM(II) BIS[HEXAFLUOROPHOSPHATE] COMPLEXES

Submitted by NICHOLAS C. THOMAS* and GLEN B. DEACON†
Checked by ANTONIO LLOBET‡ and THOMAS J. MEYER‡

The tris(2,2'-bipyridine) and tris(1,10-phenanthroline) complexes of ruthenium(II) have generated considerable interest due to their luminescence properties and capacity for electron- and energy-transfer processes. Studies of the related mixed-ligand complexes such as $[Ru(bpy)_2L]^{2+}$ and $[Ru(bpy)L_2]^{2+}$ (L = substituted bpy§ or phen§) have demonstrated that the redox properties of these complexes can be varied by altering the ligands and their substituents. Preparative routes to these mixed-ligand species generally involve the reactions of complexes containing labile ligands such as $[Ru(bpy)_2(acetone)]^{2+}$,[1] $[Ru(bpy)_2(1,2\text{-dimethoxyethane})]^{2+}$,[2] or $Ru(bpy)_2Cl_2$ (refs. 3, 4) with the desired ligand in water. The best synthesis of $[Ru(bpy)_3]Cl_2$ in aqueouis media involves the reaction of dried $RuCl_3$ (obtained from commercial $RuCl_3 \cdot xH_2O$ by careful pretreatment) with 2,2'-bipyridine and freshly prepared sodium phosphinate.[5]

The method described here uses the polymeric dicarbonyldichlororuthenium(II), $[Ru(CO)_2Cl_2]_n$, obtained quantitatively by heating at reflux a mixture of commercial $RuCl_3 \cdot xH_2O$ in formic acid,[6] which in methanol[7] or 2-methoxyethanol (this work) reacts with 2,2'-bipyridine to give $Ru(CO)_2Cl_2(bpy)$ in high yield. Treatment of the latter compound with 1,10-phenanthroline and excess trimethylamine-N-oxide (Me_3NO) in 2-methoxyethanol gives $[Ru(bpy)(phen)_2]Cl_2$, which is conveniently isolated as the hexafluorophosphate salt.[8] By a similar procedure $[Ru(bpy)_3]$ $[PF_6]_2$ can be prepared directly from $[Ru(CO)_2Cl_2]_n$. This synthesis provides an alternative nonaqueous route to tris(bidentate)ruthenium(II) compounds with the advantages of good yields and short overall reaction times.

Procedure

■ **Caution.** *Anhydrous trimethylamine-N-oxide is obtained by vacuum sublimation of the commercial dihydrate (Aldrich Chemical Company)*

*Department of Chemistry, Auburn University at Montgomery, Montgomery, AL 36193.
†Department of Chemistry, Monash University, Clayton, Victoria 3168 Australia.
‡Department of Chemistry, The University of North Carolina, Chapel Hill, NC 27514.
§bpy = 2,2'-bipyridine; phen = 1,10-phenanthroline.

at about 120°. Only small quantities of Me_3NO (~0.5–1.0 g) should be sublimed at one time and the sublimation should be conducted behind protective shielding.* All reactions employing Me_3NO should be performed in a well-ventilated fumehood since trimethylamine (Me_3N) is formed in these reactions and is inflammable, corrosive, and has the smell of rotten fish.

A. (2,2'-BIPYRIDINE)BIS(1,10-PHENANTHROLINE)-RUTHENIUIM(II) BIS(HEXAFLUOROPHOSPHATE)

$$[Ru(CO)_2Cl_2]_n + n(bpy) \xrightarrow{\text{2-methoxyethanol}} n[Ru(CO)_2Cl_2(bpy)]$$

$$[Ru(CO)_2Cl_2(bpy)] + 2(phen) + 2Me_3NO \xrightarrow{\text{2-methoxyethanol}}$$
$$[Ru(bpy)(phen)_2]Cl_2 + 2Me_3N + 2CO_2$$

$$[Ru(bpy)(phen)_2]Cl_2 + 2NH_4PF_6 \longrightarrow [Ru(bpy)(phen)_2][PF_6]_2 + 2NH_4Cl$$

A suspension of polymeric dicarbonyldichlororuthenium(II) (1.00 g, 4.4 mmol) in 2-methoxyethanol (80 mL) is heated in a 125-mL Erlenmeyer flask for several minutes until the solid dissolves. The solution is then filtered, 2,2'-bipyridine (0.80 g, 5.1 mmol) in 2-methoxyethanol (20 mL) is added, and the resulting mixture boiled in air for 5 min. After cooling the flask to 0° the contents are filtered to give crude $[Ru(CO)_2Cl_2(bpy)]$. A further portion of crude material can be obtained by concentration of the filtrate. Recrystallization from $MeOH$–CH_2CH_2 gives yellow needles. Yield: 1.32 g (80%), mp > 250°.

Anal. Calcd. for $C_{12}H_8Cl_2N_2O_2Ru$: C, 37.5; H, 2.1; Cl, 18.5; N, 7.3. Found: C, 37.9; H, 2.2; Cl, 18.5; N, 7.4%.

The ^1H NMR,[7] IR,[7] mass,[7] and UV/vis[9] spectra have been reported. This procedure may be used to prepare analogous complexes containing related bidentate ligands (e.g., 1,10-phenanthroline, 2,2'-biquinoline, 2,2'-iminodipyridine, 1,2-ethanediamine).[7,10]

A mixture of $Ru(CO)_2Cl_2(bpy)$ (0.20 g, 0.52 mmol), 1,10-phenanthroline (0.20 g, 1.1 mmol), and anhydrous trimethylamine-*N*-oxide (~0.2 g, 2.5 mmol) is heated under reflux in degassed 2-methoxyethanol (25 mL) for 2 hr under nitrogen. During this period the solution darkens and finally remains a deep red color. At this point, if the dichloride complex is desired, the solution is cooled to 0° and $[Ru(bpy)(phen)_2]Cl_2$ collected by filtration.

---

*The sublimation has been carried out over a dozen times without any adverse effects. However, one of the referees experienced an explosion while drying Me_3NO in diethyl ether.

To obtain the hexafluorophosphate salt aqueous NH_4PF_6 or KPF_6 (0.4 g in 10 mL water) is added prior to cooling. The solution is then evaporated to dryness under reduced pressure at $\sim90°$ and the residue is recrystallized from ethanol to give orange-red crystals. The yield is 0.36 g (75%).

Anal. Calc. for $C_{32}H_{24}F_{12}N_6P_2Ru$: C, 45.0; H, 2.7; N, 9.3. Found: C, 44.8; H, 2.7; N, 9.2%.

This method may also be used to prepare $[Ru(bpy)_2(phen)][PF_6]_2$ from $[Ru(CO)_2Cl_2(phen)]$ and 2,2'-bipyridine.[8]

B. TRIS(2,2'-BIPYRIDINE)RUTHENIUIM(II) BIS(HEXAFLUOROPHOSPHATE)

$$[Ru(CO)_2Cl_2]_n + 3n(bpy) + 2n(Me_3NO) \xrightarrow{\text{2-methoxyethanol}}$$
$$n[Ru(bpy)_3]Cl_2 + 2n(Me_3N) + 2n(CO_2)$$

A mixture of polymeric dicarbonyldichlororuthenium(II) (0.15 g, 0.64 mmol), 2,2'-bipyridine (0.35 g, 2.2 mmol) and anhydrous trimethylamine-N-oxide (~0.2 g, 2.5 mmol) is heated under reflux in degassed 2-methoxyethanol for 2 h under nitrogen. Work-up as described above gives $[Ru(bpy)_3][PF_6]_2$. Yield: 0.44 g (70%).

Properties

The absorption spectrum of $[Ru(bpy)(phen)_2]^{2+}$ in acetonitrile shows a maxima at 448 nm (ϵ 1.65 $\times$ 10^4) and 262 nm (ϵ 9.17 $\times$ 10^4), which have been assigned to metal-to-ligand charge transfer and $\pi \rightarrow \pi^*$ transitions, respectively.[11*] In addition shoulders at 430 and 284 nm are observed. The luminescence spectrum[12] and emission life time[13] in aqueous solution at 298 K have also been determined. Electrochemical studies estimate $E_{1/2}$ for the Ru^{3+}/Ru^{2+} couple in acetonitrile at 1.30 V.[11] The characteristic ^{1}H NMR spectrum has also been recorded.[13] The spectroscopic properties of $[Ru(bpy)_3]^{2+}$ have been summarized recently in this series.[5]

References

1. B. P. Sullivan, D. J. Salmon, and T. J. Meyer, *Inorg. Chem.*, **17**, 3334 (1978).
2. J. A. Connor, T. J. Meyer, and B. P. Sullivan, *Inorg. Chem.*, **18**, 1388 (1979).
3. P. Belser and A. von Zelewsky, *Helv. Chim. Acta*, **63**, 1675 (1980).

*The checkers report the following absorption spectrum: 453, 430 (sh), 294 (sh), and 273 nm.

4. B. Bosnich, *Inorg. Chem.*, **7**, 2379 (1968).
5. J. A. Broomhead and C. G. Young, *Inorg. Synth.*, **21**, 127 (1982).
6. M. J. Cleare and W. P. Griffith, *J. Chem. Soc. (A)*, **1969**, 372.
7. D. St. C. Black, G. B. Deacon, and N. C. Thomas, *Aust. J. Chem.*, **35**, 2445 (1982).
8. D. St. C. Black, G. B. Deacon, and N. C. Thomas, *Inorg. Chim. Acta.*, **65**, L75 (1982).
9. J. M. Kelly, C. M. O'Connell, and J. G. Vos, *Inorg. Chim. Acta*, **64**, L75 (1982).
10. N. C. Thomas and G. B. Deacon, *Synth. React. Metal Org. Inorg. Chem.*, **16**, 85 (1986).
11. R. J. Staniewicz, R. F. Sympson, and D. C. Hendricker, *Inorg. Chem.*, **16**, 2166 (1977).
12. G. A. Crosby and W. H. Elfring, *J. Phys. Chem.*, **80**, 2206 (1976).
13. J. E. Baggot, G. K. Gregory, M. J. Pilling, S. Anderson, and K. R. Sneddon, *J. Chem. Soc. Faraday*, **2**, 79, 195 (1983).

24. POTASSIUM TRIALKYL- AND TRIARYLSTANNATES: PREPARATION BY THE DEPROTONATION OF STANNANES WITH POTASSIUM HYDRIDE

$$R_3SnH + KH \longrightarrow K[SnR_3] + H_2$$
$$(R = n\text{-Bu,Ph})$$

$$K[SnR_3] + D_2O \longrightarrow R_3SnD + KOD$$

$$K[SnR_3] + n\text{-BuBr} \longrightarrow R_3Sn(n\text{-Bu}) + KBr$$
$$(R = n\text{-Bu})$$

Submitted by R. CORRIU,* C. GUERIN,* and B. KOLANI*
Checked by M. NEWCOMB† and M. T. BLANDA†

The reaction of alkyl halides with alkali trialkyl- and triarylstannates (Li, Na) has received great attention[1] as a method of formation of the Sn—C bond. However, debated data were reported in the literature[2]; they apparently depend on the nature of the anionic species and on the experimental conditions (temperature, solvent, additives). Moreover, it was reported that the decomposition of stannate anions is very sensitive to additives and accelerated by R_3Sn—SnR_3.[3]

As the presence of salts and by-products is a factor of importance, we report here a clean preparation of potassium stannate reagents by the

*Institut de Chimie Fine, U.A. CNRS 1097, U.S.T.L., Place E. Bataillon, 34060 Montpellier Cedex, France.
†Department of Chemistry, Texas A&M, College Station, TX 77843.

deprotonation of trialkyl- or triarylstannanes, R_3SnH, with potassium hydride [eq. (1)].

$$R_3SnH + KH \longrightarrow K[SnR_3] + H_2 \tag{1}$$

This method can be applied either to aryl (R = Ph) or alkyl (R = n-Bu) stannanes in a solvent (S) such as diethyl ether (Et_2O), tetrahydrofuran (THF), 1,2-dimethoxyethane (DME), or hexane.

A. POTASSIUM TRIPHENYLSTANNATE

$$Ph_3SnH + KH \xrightarrow{\text{S}} K[SnPh_3] + H_2 \tag{2}$$
$$\text{(S = hexane, Et}_2\text{O, THF, DME)}$$

Materials

Triphenylstannane is prepared by reduction of the triphenyltin chloride with $LiAlH_4$ in diethyl ether at room temperature, as described previously.[4]

Diethyl ether or hexane are dried by heating them, refluxing, and subsequent distillation over sodium wire; THF over CaH_2 and DME over $LiAlH_4$ immediately before use. Solvents are degassified under vacuum. Potassium hydride (20% in oil)* is cautiously washed before use by shaking it three times with anhydrous hexane. The resulting solid residue is subjected to pumping under vacuum to remove any residual solvent.

Schlenck-type glassware is used in all procedures.[5] All manipulations are carried out under dry and oxygen-free nitrogen, using conventional vacuum line techniques.[5]

■ **Caution.** *Dry potassium hydride is a dangerous compound from which water and air must be excluded. All manipulations of this compound must be carried out in a dry oxygen-free nitrogen-purged atmosphere.*

Stannanes are toxic and must be handled in an efficient hood at all times.[6]

Procedure

The procedure is described only for the case of 1,2-dimethoxyethane as solvent; similar experimental conditions are used in the case of THF (1 hr, rt), diethyl ether (6 hr, rt), and hexane (6 hr, 60°).

---

*Potassium hydride is purchased from Fluka A.G., CH-9470 Buchs (20 wt% in oil) or from Aldrich Chemical Company, Milwaukee, WI 53233 (35 wt% in oil).

A suspension of 400 mg (10 mmol) of potassium hydride, cautiously washed with dry hexane, in 10 mL of 1,2-dimethoxyethane is stirred magnetically at room temperature under a nitrogen atmosphere in a 50-mL Schlenck tube. A solution of 3.5 g (10 mmol) of triphenylstannane in 15 mL of 1,2-dimethoxyethane is then added dropwise via a pressure-equalized dropping funnel. Evolution of hydrogen is observed, and a green-yellow color develops immediately. The reaction is monitored either by IR or ^{1}H NMR spectroscopy using deuterolyzed aliquots of the reaction mixture obtained as follows: a 1.0- to 2.0-mL sample of the solution to be tested is cautiously added via a syringe to 5.0 mL of cooled D_2O. After a few minutes, the organic layer is extracted with 10 mL of ether, dried over anhydrous magnesium sulfate, and freed of solvent under vacuum. The IR spectrum in benzene of the residue shows the gradual disappearance of a strong band at 1825 cm^{-1} (ν_{Sn-H}), while a characteristic absorption at 1323 cm^{-1} (ν_{Sn-D}) appears. The ^{1}H NMR spectrum shows the gradual disappearance of the Sn—H singlet at $\delta = 5.3$ ppm (C_6D_6). After ~25 min at room temperature, the reaction is complete; no ν_{Sn-H} can be detected and hence Ph$_3$SnH is <0.5%. The deprotonation of Ph$_3$SnH was run for 1 hr in THF, 6 hr in Et$_2$O, and 6 hr in hexane (46% yield determined by ^{1}H NMR spectroscopy). The potassium triphenylstannate was characterized as its corresponding triphenyltin deuteride[7] [Ph$_3$SnD: 90% yield, bp 152–156° (0.002 torr)].

B. POTASSIUM TRIBUTYLSTANNATE

$$(n\text{-Bu})_3\text{SnH} + \text{KH} \xrightarrow{\text{S}} \text{K}[\text{Sn}(n\text{-Bu})_3] + \text{H}_2 \qquad (3)$$
$$\textbf{(S = THF, DME)}$$

Materials

Tributylstannane is prepared according to a published method, via an exchange between bis(tributyl)oxide and a polysiloxane containing Si—H bonds.[8]

Procedure

The procedure is described in dimethoxyethane as solvent: similar experimental conditions are used in THF.

A suspension of 400 mg (10 mmol) of potassium hydride, cautiously washed with dry hexane, in 10 mL of dimethoxyethane is stirred magnetically at room temperature under a nitrogen atmosphere in a 50-ml Schlenck tube. A solution of 2.9 g (10 mmol) of tributylstannane in 15 mL of dimethoxyethane is added dropwise via a pressure-equalized dropping funnel. Evolution of hydrogen is observed and a yellow-green color develops after 10 min. The reaction mixture is then stirred at room temperature. The reaction can be monitored either by $IR[(n\text{-Bu})_3SnH$: benzene, $\nu_{Sn—H} = 1807 \text{ cm}^{-1}]$ or ^1H NMR spectroscopy $[(n\text{-Bu})_3Sn—H$: C_6D_6, $\delta = 4.78$, m, 1 H (Sn—H)] as described in the case of potassium triphenylstannate. After ~30 min at room temperature, the reaction is complete. The deprotonation of $(n\text{-Bu})_3SnH$ was run in THF for 1 hr.* The potassium tributylstannate was characterized as its corresponding tributyltin deuteride[9] [$n\text{-Bu}_3SnD$: 85% yield, bp 50–55° (0.001 torr)].

C. PROPERTIES†

Potassium trialkyl- and triarylstannates are air and moisture sensitive. They are thermally instable, leading to $R_3Sn—SnR_3$, which is known to catalyse the decomposition of stannyl anions.[3] They must be handled and stored in an inert atmosphere (nitrogen or argon) by normal vacuum line techniques. For a prolonged storage, it is desirable to store the substance at low temperatures (below −40°).

Water and alcohol react quantitatively with potassium stannates, leading to the corresponding triaryl- or trialkylstannanes [Ph_3SnH: 95% yield, bp 165–168° (0.3 torr), mp 28°; $(n\text{-Bu})_3SnH$: 90% yield, bp 68–74° (0.3 torr)]. Deuterolysis is well suited for determination of stannyl anion content.

References

1. J. P. Quintard and M. Pereyre, *Rev. Silicon, Germanium, Tin and Lead Compounds,* **4,** 153 (1980); M. Pereyre, J. P. Quintard, and A. Rahm, *Tin in Organic Synthesis,* Butterworths, London, 1987.
2. For instance: G. F. Smith, H. Kuivila, R. Simon, and L. J. Sultan, *J. Am. Chem. Soc.,* **103,** 833 (1981); J. San Filippo Jr., and J. Silbermann, *ibid.,* **104,** 2831 (1982); W. Kitching, H. Olszowy, and K. J. Hawey, *J. Org. Chem.,* **47,** 1893 (1982); M. Newscomb and A. R. Courtney, *J. Org. Chem.,* **45,** 1807 (1980); M. Newcomb and H. G. Smith, *J. Organometal.*

*The checkers report a longer reaction time, that is, 2 hr, for the deprotonation of Bu_3SnH in THF.

†The checkers report experiments in which they alkylated the potassium tributylstannate and isolated and characterized the products as a measure of the yield of the reactions: (a) Ph_3SnK was alkylated at 25° by addition of BuBr (1.37 g, 10 mmol). After 4 hr at 25°, the reaction mixture was diluted with 20 mL of THF and filtered. Distillation of the solvent at reduced pressure gave a solid, which was recrystallized from hexane to give 3.37 g (83% yield of Ph_3SnBu, mp 61–63°). (b) Bu_3SnK was alkylated at 25° by addition of BuBr (1.37 g, 10 mmol). After 4 hr at 25°, the reaction mixture was diluted with 20 mL of THF and filtered. Distillation of the solvent at reduced pressure gave a liquid that was distilled to give 2.67 g (77% yield) of Bu_4Sn, bp 108–110° (5 torr).

Chem., **228**, 61 (1982); E. C. Ashby and R. N. De Priest, *J. Am. Chem. Soc.*, **104**, 6144 (1982); K. W. Lee and J. San Filippo Jr., *Organometallics*, **2**, 906 (1983); E. C. Ashby, R. N. De Priest, and W. Y. Su, *Organometallics*, **3**, 1718 (1984), and references therein.
3. K. Kobayashi, H. Kawanisi, T. Hitomi and S. Kozima, *J. Organometal. Chem.*, **233**, 299 (1982).
4. H. Gilman and J. Eisch, *J. Org. Chem.*, **20**, 763 (1955); H. G. Kuivila and O. F. Beumel Jr., *J. Am. Chem. Soc.*, **83**, 1246 (1961); H. G. Kuivila, *Synthesis*, **1970**, 499; J. J. Eisch, *Organometallic Synthesis*, Vol. 2, J. J. Eisch and R. B. King (eds.), Academic Press, New York, 1981. p. 173.
5. D. F. Shriver, *The Manipulation of Air-Sensitive Compounds*, McGraw-Hill, New York, 1969.
6. R. C. Poller, *Chemistry of Organotin Compounds*, Logos Press, London, 1970.
7. W. P. Neumann and R. Sommer, *Angew Chem. Int. Ed. (Engl.)*, **2**, 547 (1963).
8. K. Hayashi, J. Iyoda, and I. Shiihara, *J. Organometal. Chem.*, **10**, 81 (1967).
9. H. J. Albert and W. P. Neumann, *Synthesis*, **1980**, 942.

25. (BENZENETHIOLATO)TRIBUTYLTIN

$$[(n\text{-}C_4H_9)_3Sn]_2O + 2C_6H_5SH \longrightarrow 2Sn(SC_6H_5)(n\text{-}C_4H_9)_3 + H_2O$$

Submitted by PAUL M. TREICHEL* and MARVIN H. TEGEN*
Checked by STEPHEN A. KOCH†

Organotin thiolate complexes such as the title compound are useful reagents for the synthesis of transition metal complexes with RS⁻ ligands.[1] Examples of this use are given in the procedures that follow this synthesis.

It is possible to prepare compounds of the general formula $Sn(SR')R_3$ from reactions of a mercaptan ($R'SH$) and either $Sn(hal)R_3$ (hal = Cl, Br, I), $Sn(OH)R_3$, or $(R_3Sn)_2O$ (R,R' = alkyl, aryl groups).[2,3] The example given here uses μ-oxo-bis(tributyltin) as the organotin precursor because this is one of the least expensive tin compounds available.

Procedure

■ **Caution.** *Organotin compounds are very toxic and should be handled accordingly.*

A 29.8 g (0.050 mol) sample of $[(n\text{-}C_4H_9)_3Sn]_2O$‡ is placed in a 100 mL, three-neck, round-bottom flask equipped with magnetic stirrer, condenser, dropping funnel, and nitrogen inlet. From the dropping funnel, 11.0 mL (0.107 mol) of C_6H_5SH is added slowly. A reaction occurs immediately as

*Department of Chemistry, University of Wisconsin, Madison, WI 53706.
†Department of Chemistry, State University of New York, Stony Brook, NY 11794.
‡Aldrich Chemical Co., P. O. Box 355, Milwaukee, WI 53201.

evidenced by heat evolution and formation of a white emulsion. The reaction flask is attached to a short distillation column and the mixture heated to boiling. Over a period of about 15 min the water formed in the reaction is distilled from the system, and the solution becomes clear. After water has ceased to distill, the system is cooled and then distillation resumed *in vacuo*. A small amount of benzenethiol distills first at a relatively low temperature. The product then distills at 172 to 174°/2.5 torr. Isolated yield: ~36 g, >90% yield. It can be further purified by redistillation if desired.

Properties

(Benzenethiolato)tributyltin is a clear, viscous liquid. It is thermally stable and unreactive toward either oxygen or water under ambient conditions. This procedure can be used to prepare many other similar compounds including $Sn(SCH_3)(n\text{-}C_4H_9)_3$ (bp 118–122°, 0.8 torr) and $Sn[SC(CH_3)_3](n\text{-}C_4H_9)_3$ (bp 138–143°, 0.8 torr).

It is reported that $(R_3Sn)_2O$ compounds react with acids (HX) that have ionization constants between 10^{-1} and 10^{-10} to produce $SnXR_3$ species.[2] Thus, the preparation of $Sn(SPh)(n\text{-}C_4H_9)_3$ is an example of a more general synthetic procedure.

References

1. The first report of this type of reaction is E. W. Abel, B. C. Crosse, and D. B. Brady, *J. Am. Chem. Soc.*, **87**, 4395 (1965). Additional references are included in the following article.
2. G. S. Sasin, *J. Org. Chem.*, **18**, 1142 (1953).
3. E. W. Abel and D. B. Brady, *J. Chem. Soc.*, **1965**, 1192.

26. USE OF (BENZENETHIOLATO)TRIBUTYLTIN TO PREPARE COMPLEXES OF MANGANESE CARBONYL HAVING BRIDGING THIOLATE LIGANDS

Submitted by PAUL M. TREICHEL* and MARVIN H. TEGEN*
Checked by STEPHEN A. KOCH†

Thiolate anions, RS⁻ (R = alkyl, aryl groups), may coordinate to a single metal atom as a monodentate ligand but, more commonly, these species function as bridging ligands between two or three metal atoms. Many

*Department of Chemistry, University of Wisconsin, Madison, WI 53706.
†Department of Chemistry, State University of New York, Stony Brook, NY 11794.

organometallic species with such ligands are known including several whose preparations are reported in previous volumes of *Inorganic Syntheses*.[1] Interest in metal complexes with thiolate ligands has increased in recent years because of the involvement of such species in biological processes and in the area of catalysis.[2]

Preparations of three manganese carbonyl thiolate complexes are described here. The procedures use (benzenethiolato)tributyltin, $Sn(SC_6H_5)(n\text{-}C_4H_9)_3$, whose preparation was given in the previous example, as a source of the thiolate ligand. The reactions are simple to carry out and provide a high yield of the desired product. The given procedures are also generally applicable to preparations of manganese complexes with other thiolate groups. Preparations of thiolate derivatives of other metals using organotin thiolate reagents are known.[3]

A. BIS-μ-(BENZENETHIOLATO)-OCTACARBONYLDIMANGANESE(I)

$$2MnBr(CO)_5 + 2Sn(SC_6H_5)(n\text{-}C_4H_9)_3 \longrightarrow$$
$$Mn_2(\mu\text{-}SC_6H_5)_2(CO)_8 + 2SnBr(n\text{-}C_4H_9)_3 + 2CO$$

Procedure

■ **Caution.** *Carbon monoxide and metal carbonyl compounds are toxic and all reactions must be carried out in an efficient fume hood. The organotin compound is also toxic.*

A 1.2-mL sample of $Sn(SC_6H_5)(n\text{-}C_4H_9)_3$ (1.40 g, 3.5 mmol) is added to a suspension of 0.81 g (1.64 mmol) $Mn_2(\mu\text{-}Br)_2(CO)_8$ (ref. 4)* in 30 mL of freshly distilled THF in a 100-mL flask equipped with N_2 inlet, stirrer, and dropping funnel. This mixture is stirred at ambient temperature under nitrogen for 1 hr; then the volume of solvent is reduced by evaporation in vacuum to ~3 mL. Cooling this solution to $-15°$ causes orange crystals of the product to precipitate. These crystals are collected by filtration, washed twice with cold (0°) hexane, and allowed to dry in air. Yield: 0.85 g, 94%. Recrystallization of the material can be carried out using hexane as a solvent.

---

*The use of $Mn_2(\mu\text{-}Br)_2(CO)_8$ rather than $MnBr(CO)_5$ in this reaction is recommended. Bromomanganese pentacarbonyl also reacts with $Sn(SC_6H_5)(n\text{-}C_4H_9)_3$ to give the desired product; however, more forcing conditions are required, which result in concurrent formation of $Mn_4(\mu\text{-}SC_6H_5)_4(CO)_{12}$.

Properties

Orange, crystalline $Mn_2(\mu\text{-}SC_6H_5)_2(CO)_8$ decomposes without melting at 154°. The compound may be stored, as a solid, for prolonged periods without substantial thermal decomposition. It is not particularly sensitive to either atmospheric oxygen or to water. Facile conversion to $Mn_4(\mu\text{-}SC_6H_5)_4(CO)_{12}$ occurs upon heating; a convenient procedure involves heating a solution of $Mn_2(\mu\text{-}SC_6H_5)_2(CO)_8$ in THF at reflux for several hours. Because of the ease of conversion of dimer to tetramer, the latter species may appear as a contaminant in synthesis of various $Mn_2(\mu\text{-}SR)_2(CO)_8$ compounds. It is easy to detect the presence of the tetramer using IR spectroscopy by the appearance of ν_{CO} absorptions at about 2020 (s) and 1950 (m)cm^{-1} [2021 (s) and 1951 (m)cm^{-1} for $Mn_4(\mu\text{-}SC_6H_5)_4(CO)_{12}$ in hexane]. Infrared absorptions (ν_{CO}) for $Mn_2(\mu\text{-}SC_6H_5)_2(CO)_8$ occur at 2084 (m), 2028 (s), 2022 (s), 2005 (m), 1980 (s), 1965 (m)cm^{-1} (hexane).

Conversion of several $Mn_2(\mu\text{-}SR)_2(CO)_8$ complexes [R = CH_3, H, $Sn(CH_3)_3$] to the $Mn(SR)(CO)_5$ species is known to occur under pressure of CO.[8] These monomeric species are unstable, rapidly reverting back to $Mn_2(\mu\text{-}SR)_2(CO)_8$ when the CO pressure is released.

Preparation of analogous complexes of $Mn_2(\mu\text{-}SR)_2(CO)_8$ [e.g., R = $C(CH_3)_3$, mp 146°(dec); R = CH_3, mp 123°(dec)] can be carried out by this procedure. When the given procedure is applied to other systems, however, it is advisable to monitor ν_{CO} values for the reaction mixture to ascertain optimum conditions for the specific product. Both duration of time and temperature of the reaction are influential in determining the ratio of $Mn_2(\mu\text{-}SR)_2(CO)_8$ to $Mn_4(\mu\text{-}SR)_4(CO)_{12}$ in these reactions.

This compound has also been prepared by the reaction of $MnX(CO)_5$ (X = Cl, Br) and C_6H_5SH;[5,6] related complexes $Mn_2(\mu\text{-}SR)_2(CO)_8$ (R = CH_3, C_2H_5, $n\text{-}C_4H_9$) have been made by reaction of $MnH(CO)_5$ and RSSR.[7] The yield of product by the former route is not good (and the procedure has not been widely used) because it is difficult to set conditions so that $Mn_4(\mu\text{-}SPh)_4(CO)_{12}$ is not obtained concurrently. The latter procedure requires a reagent, $MnH(CO)_5$, which is air sensitive and less easily handled.

B. TETRA-μ_3-(BENZENETHIOLATO)-DODECACARBONYLTETRAMANGANESE(I)

$$4MnBr(CO)_5 \;+\; 4Sn(SC_6H_5)(C_4H_9)_3 \longrightarrow$$
$$Mn_4(\mu\text{-}SC_6H_5)_4(CO)_{12} \;+\; 4SnBr(n\text{-}C_4H_9)_3 \;+\; 8CO$$

Procedure

■ **Caution.** *Carbon monoxide and metal carbonyl compounds are toxic and all reactions must be carried out in an efficient fume hood. The organotin compound is also toxic.*

A solution of 2.9 g (7.3 mmol) $MnBr(CO)_5$ and 2.9 mL (3.4 g, 8.4 mmol) $Sn(SC_6H_5)(n\text{-}C_4H_9)_3$ in 40 mL of THF in a 100-mL round-bottom flask (N_2 inlet, reflux condenser) is heated at reflux for 6 hr. The solution is allowed to cool. Solvent volume is reduced to 5 mL in vacuum; cooling at $-15°$ causes the orange crystalline product to precipitate from solution. This species is separated by filtration and air dried, yield 1.7 g, 93%. Recrystallization can be carried out using hexane.

Properties

The product is thermally stable and does not react with atmospheric oxygen or water. It decomposes without melting at 228° and it has ν_{CO} absorptions at 2021 (s), 1951 (m) cm^{-1}. This tetrameric species has metal and sulfur atoms at alternate corners of a cube.[9]

Related compounds [e.g., $Mn_4(\mu\text{-}SR)_4(CO)_{12}$ (R = CH_3, mp 207° dec.)] may be prepared in similar reactions. As noted, it is possible to convert the $Mn_2(\mu\text{-}SR)_2(CO)_8$ species to these compounds by heating in THF, carbon monoxide being evolved. Preparation of the title compound has also been accomplished by reaction of $MnBr(CO)_5$ and C_6H_5SH; a low yield was reported.[10]

C. TETRAETHYLAMMONIUM TRIS-(μ-BENZENETHIOLATO)-HEXACARBONYLDIMANGANATE(I)[11]

$$(C_2H_5)_4N[Mn_2(\mu\text{-}Br)_3(CO)_6] + 3Sn(SC_6H_5)(n\text{-}C_4H_9)_3 \longrightarrow$$
$$(C_2H_5)_4N[Mn_2(\mu\text{-}SC_6H_5)_3(CO)_6] + 3SnBr(n\text{-}C_4H_9)_3$$

Procedure

■ **Caution.** *Carbon monoxide and metal carbonyl compounds are toxic and all reactions must be carried out in an efficient fume hood. The organotin compound is also toxic.*

Samples of $(C_2H_5N)_4N[Mn_2(\mu\text{-}Br)_3(CO)_6]$ (ref. 12) (1.60 g, 2.5 mmol) and $Sn(SC_6H_5)(n\text{-}C_4H_9)_3$ (3.4 mL, 4.0 g, 10.0 mmol) are dissolved in 40 mL of anhydrous methanol contained in a 100-mL round-bottom flask (N_2 inlet, reflux condenser). The solution is heated at reflux for 1 hr, then allowed to cool. The volume of solvent is reduced to about 10 mL *in vacuo*. This solution is then cooled at $-20°$ for several hours, an orange solid precipitating during this time. This solid is collected by filtration and dried *in vacuo*. It is then dissolved in a minimum volume of CH_2Cl_2, and following filtration an equal volume of diethyl ether is added. Cooling this solution causes precipitation of the crystalline product that is collected by filtration and dried *in vacuo;* Yield: 1.70 g (92%).

Anal. Calcd. for $C_{32}H_{35}NO_6S_3Mn_2$: C, 52.24; H, 4.80%. Found: C 52.52; H, 4.77%.

Properties

The compound is an orange crystalline species, mp 192°, which may be stored in a screw-cap vial for a prolonged period of time without decomposition or reaction with air or moisture. It has a characteristic IR absorption pattern; ν_{CO} at 1997 (s), 1918 (s), 1908 (sh) cm^{-1} in CH_2Cl_2. It is known to react with electrophiles (CF_3CO_2H, $[(CH_3)_3O]BF_4$) in the absence of additional ligands to give $Mn_4(\mu_3\text{-}SC_6H_5)_4(CO)_{12}$. In the presence of additional ligands, either $Mn_2(\mu\text{-}SC_6H_5)_2(CO)_6(L)_2$ [L = CO, $P(CH_3)_3$] or $Mn_2(\mu\text{-}SC_6H_5)_2(\mu\text{-}CO)(CO)_4(L)_2$ [L = $P(C_6H_5)_3$] is formed.[11]

References

1. $Ir_2(\mu\text{-}S\text{-}t\text{-}C_4H_9)_2(CO)_4$: D. de Montauzon and R. Poilblanc, *Inorg. Synth.*, **20,** 237 (1980); $Fe_2(\mu\text{-}SC_2H_5)_2(CO)_2(\eta\text{-}C_5H_5)_2$: G. J. Kubas and P. J. Vergamini, *Inorg. Synth.*, **21,** 37 (1982).
2. G. C. Kuehn and S. S. Isied, *Progress in Inorganic Chemistry*, Vol. 27 S. J. Lippard, (ed.), Wiley-Interscience, New York, 1980, p. 154.
3. See examples in W. Ehrl and H. Vahrenkamp, *Chem. Ber.*, **105,** 1471 (1972).
4. F. Calderazzo, R. Poli, and D. Vitale, *Inorg. Synth.*, **23,** 32 (1985).
5. W. Hieber and W. K. Schropp, *Z. Naturforsch.*, **14B,** 460 (1959).
6. M. Ahmad, G. R. Knox, F. J. Preston, and R. I. Reed, *Chem. Commun.*, **1967,** 138.
7. P. M. Treichel, J. H. Morris, and F. G. A. Stone, *J. Chem. Soc.*, **1963,** 720.
8. V. Kullmer and H. Vahrenkamp, *Chem. Ber.*, **109,** 1569 (1976).
9. B. F. G. Johnson, P. J. Pollick, I. G. Williams, and A. Wojcicki, *Inorg. Chem.*, **7,** 831 (1968).
10. A. G. Osborne and F. G. A. Stone, *J. Chem. Soc.*, **1966,** 1143.
11. P. M. Treichel and M. H. Tegen, *J. Organomet. Chem.*, **242,** 385 (1985).
12. B. J. Brisdon, D. A. Edwards, and J. W. White, *J. Organomet. Chem.*, **161,** 233 (1978).

27. METHYLENEBIS[DICHLOROPHOSPHINE],* CHLOROBIS[(DICHLOROPHOSPHINO)METHYL]- PHOSPHINE,† AND METHYLENEBIS[DIMETHYL PHOSPHINE]

Submitted by S. HIETKAMP,‡ H. SOMMER,‡ and O. STELZER‡
Checked by A. L. BALCH,§ J. C. LINEHAN,§ and D. E. ORAM§

Phosphine ligands with the P—C—P donor sequence are of interest since they may bridge metal–metal bonds and thus stabilize oligometallic or cluster compounds.[1,2]

Methylenebis[dichlorophosphine] has been used in the syntheses of a variety of bidentate ligands with a P—C—P skeleton.[3,4]

Methylenebis[dichlorophosphine], Cl_2P—CH_2—PCl_2, has originally been reported by Sommer[5] and Fild et al.[5a] Later, Novikova et al.,[6] published its synthesis using CH_2Cl_2, Al, and PCl_3 as starting materials. Aluminum and dichloromethane react to give organoaluminum compounds, Cl_2Al—CH_2—$AlCl_2$[7] or Cl_2Al—$[CH_2$—$AlCl]_n$—CH_2—$AlCl_2$,[7] which with phosphorus trichloride yield the methylenebis[dichlorophosphine]. While studying this reaction in order to increase the yield for Cl_2P—CH_2—PCl_2 we found that, in addition to the desired product, a tridentate chloro-phosphine, Cl_2P—CH_2—PCl—CH_2—PCl_2 had been formed.[8] The compounds Cl_2P—CH_2—PCl_2 and Cl_2P—CH_2—PCl—CH_2—PCl_2 are suitable starting materials for the preparation of Me_2P—CH_2—PMe_2 and Me_2P—CH_2—PMe—CH_2—PMe_2,[9] respectively.

A. PREPARATION OF THE ORGANOALUMINUM INTERMEDIATES

In a 2000-mL three-necked flask fitted with a reflux condenser, a 500-mL pressure equalizing dropping funnel, and a mechanical stirrer, 107.9 g (4 mol) aluminum granules‖ were heated with 20 mL of dibromomethane and 120 mL of dichloromethane to 40° until reflux begins. After the reaction begins, a further 880 mL of dichloromethane are added within 1 hr. The

---

*Methylenebis[phosphorous dichloride]

†Chlorophosphinidenebis(methylene)bis[phosphorous dichloride]

‡Fachbereich 9, Anorganische Chemie, Bergische Universität-GH Wuppertal, Gauss-strasse 20, D-5600 Wuppertal 1, Federal Republic of Germany.

§Department of Chemistry, University of California, Davis, CA 95616.

‖The aluminum granules (purchased from Riedel de Haen) were heated *in vacuo* at 100 to 150° for 12 hr.

reaction mixture is heated at reflux until all the aluminum has dissolved (~90 hr). The ^{1}H − NMR spectrum of this solution shows a sharp signal at − 0.37 ppm and a broad band at − 0.2 ppm.

B. METHYLENEBIS[DICHLOROPHOSPHINE] AND CHLOROBIS[(DICHLOROPHOSPHINO)-METHYL]PHOSPHINE

To the solution of the organoaluminum compounds, 549.3 g (4 mol) of phosphorus trichloride, diluted with 0.8 L of dichloromethane, is added dropwise over a period of 3 hr at a rate that keeps the reaction mixture at reflux. Thereafter, 613.3 g (4 mol) of phosphoryl chloride and 298.2 g (4 mol) of potassium chloride (ground and dried at 100°) are added within 1 hr. The reaction mixture is heated at reflux for an additional hour. Most of the dichloromethane is removed by distillation at normal pressure,* and the waxy residue is extracted with four 500-mL volumes dry petroleum (bp 40/60). The petroleum is removed by distillation at normal pressure. Fractional distillation of the oily residue at 1.0 mbar, using a 20-cm Vigreux column affords methylenebis[dichlorophosphine]. Yield: 134–160 g (32–38%), bp 54°/1 mbar.

Methylenebis[dichlorophosphine] is a colorless air- and moisture- sensitive liquid. It may be characterized by its NMR spectral data: ^{1}H NMR: δ_H = 2.5 ppm [triplet, $^2J_{PH}$ = 16.0 Hz]; ^{13}C {^{1}H} NMR: δ_C = 63.9 ppm [triplet, $^1J_{PC}$ = 66.3 Hz]; ^{31}P {^{1}H} NMR: δ_P = 175.2 ppm.

From the remaining residue all volatile products are removed *in vacuo* at 50°/0.04 mbar followed by extraction with 4 × 100 mL of petroleum (bp 40/60). The petroleum and other volatile materials are stripped off *in vacuo* (0.04 mbar) at 20–50°. Almost pure (95%) chlorobis-[(dichlorophosphino)methyl]phosphine is left behind. Yield: 21–30 g (5.4–7.7%). Chlorobis[(dichlorophosphino)methyl]phosphine is a viscous air- and moisture-sensitive liquid that may be characterized by its NMR spectral data: ^{13}C {^{1}H}–NMR: δ_P = 48.6 ppm; $^1J_{^{31}P-^{13}C}$ = 52.7 Hz; $^1J_{^{31}P-^{13}C}$ = 62.3 Hz (ABCX-type spin system; A, B, C = ^{31}P; X = ^{13}C); ^{31}P {^{1}H} NMR: δ_P = 84.3 ppm (PCl); 181.2 ppm (PCl$_2$); $^2J_{^{31}P-^{31}P}$ = 69.4 Hz.

C. METHYLENEBIS[DIMETHYLPHOSPHINE]

In a 2-L three-necked flask fitted with a mechanical stirrer, reflux condenser, and a 250-mL pressure equalizing dropping funnel 67.2 g (0.31 mol) of Cl$_2$P—CH$_2$—PCl$_2$ (dissolved in 500 mL dry diethyl ether) is added

---

*The temperature in the reaction mixture should not exceed 80°.

gradually to 0.95 L of a 1.37 M solution of chloromethylmagnesium in diethyl ether at $-30°$ over a period of 3 hr with stirring. After the reaction mixture has been warmed to room temperature, it is stirred for 2 hr. Oxygen-free water ($\sim$100 mL) is added until the magnesium chloride is precipitated. The clear etheral solution is decanted and dried with Na_2SO_4, and the diethyl ether is removed by distilling off at normal pressure. Fractionated distillation of the residue at 18 mbar affords methylene-bis[dimethylphosphine]. Yield: 29.0 g (69%), bp 53–54°/18 mbar.

Methylenebis[dimethylphosphine] is a colorless, very air-sensitive liquid that spontaneously burns in contact with air. NMR-spectral data: ^1H NMR: $\delta_H = 1.4$ ppm [triplet, $^3J_{PH} = 1.0$ Hz CH_2 groups]; 1.3 ppm (multiplet, CH_3 groups); ^{13}C {^1H} NMR: $\delta_C = 36.3$ ppm [triplet, CH_2 group, $^1J_{^{31}P-^{13}C}$]; 16.7 ppm (CH_3 groups); ^{31}P {^1H} NMR: $\delta_P = -55.8$ ppm.

References

1. R. J. Puddephatt, *Chem. Soc. Rev.*, **12**, 99 (1983).
2. D. J. Brauer, S. Hietkamp, H. Sommer, and O. Stelzer, *Angew. Chem.*, **96**, 696 (1984).
3. S. Hietkamp, H. Sommer, and O. Stelzer, *Chem. Ber.*, **117**, 3400 (1984).
4. H. H. Karsch, *Z. Naturforsch.*, **38b**, 1027 (1983).
5. K. Sommer, *Z. Anorg. Allg. Chem.*, **376**, 37 (1970). (a) M. Fild, J. Heinze, and W. Krüger, *Chemiker Ztg.*, **101**, 259 (1977).
6. Z. S. Novikova, A. A. Prishchenko, and I. F. Lutsenko, *Zh. Obshch. Khim.*, **47**, 775 (1977).
7. H. Lehmkuhl and R. Schäfer, *Tetrahedron Lett.*, **21**, 2315 (1966).
8. D. J. Brauer, S. Hietkamp, H. Sommer, O. Stelzer, G. Müller, M. J. Romao, and C. Krüger, *J. Organomet. Chem.*, **296**, 411 (1985); S. Hietkamp, H. Sommer, and O. Stelzer, *Angew. Chem.*, **94**, 368 (1982).
9. H. H. Karsch, *Z. Naturforsch.*, **37b**, 284 (1982); H. H. Karsch and H. Schmidbaur, *Z. Naturforsch.*, **32b**, 762 (1977).

28. 1,4,7,10,13,16-HEXATHIACYCLOOCTADECANE (HEXATHIA-18-CROWN-6) AND RELATED CROWN THIOETHERS

$HS(CH_2)_2S(CH_2)_2SH$ + 2NaOEt + $2Cl(CH_2)_2OH \longrightarrow$

$$HOCH_2(CH_2SCH_2)_3CH_2OH$$

$HOCH_2(CH_2SCH_2)_3CH_2OH$ + $2SOCl_2 \longrightarrow$

$$ClCH_2(CH_2SCH_2)_3CH_2Cl + 2SO_2 + 2HCl$$

$ClCH_2(CH_2SCH_2)_3CH_2Cl$ + $HS(CH_2)_2S(CH_2)_2SH$ + $Cs_2CO_3 \longrightarrow$

$$\text{hexathia-18-crown-6} + 2CsCl + H_2O + CO_2$$

Submitted by ROBERT E. WOLF, JR.,* JUDITH ANN R. HARTMAN,*
LEO A. OCHRYMOWYCZ,† and STEPHEN R. COOPER*
Checked by MAHMOOD SABAHI‡ and RICHARD S. GLASS‡

Crown thioethers such as hexathia-18-crown-6 (1,4,7,10,13,16-hexathia-cyclooctadecane, 18S6)[1-6] and related ligands[7-11] form complexes with a

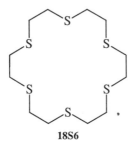

18S6

variety of transition metal ions. They have found extensive use in the synthesis of model systems for the blue copper proteins.[12-15] Previously hexathia-18-crown-6 was prepared by the reaction of ethanedithiol[16] or 2,2'-thiobis(ethanethiol)[17-18] with ethylene bromide, but both of these reactions give only low yields and require tedious work-up procedures owing to extensive formation of polymeric materials. Recent advances in the chemistry of macrocyclic sulfides by Kellogg and co-workers[19] and Ochrymowycz et al.[20] now make hexathia-18-crown-6 and related ligands readily available.

Procedure

2,2'-[Thiobis(2,1-ethanediylthio)]bis(ethanol). A 1-L three-necked flask equipped with an additional funnel, condenser, and magnetically activated stirring bar is placed in a fume hood and purged with nitrogen. Then 0.5 L of absolute ethanol is placed into it, and sodium metal (11.8 g, 0.513 mol) is slowly added in small pieces. (■ **Caution.** *The hydrogen evolved poses an explosion hazard.*) When all the sodium has undergone reaction, vacuum distilled 2,2'-thiobis(ethanethiol) (Aldrich; 39.5 g, 0.256 mol) is added under nitrogen with stirring. The resulting solution is brought to

*Inorganic Chemistry Laboratory, University of Oxford, South Parks Road, Oxford OX1 3QR, United Kingdom, and Department of Chemistry, Harvard University Cambridge, MA 02138.

†Department of Chemistry, University of Wisconsin, Eau Claire, WI 54701.

‡Department of Chemistry, University of Arizona, Tucson, AZ 85721.

reflux in an oil bath and 2-chloroethanol (34.4 mL, 0.513 mol), taken from a freshly opened bottle, is carefully added dropwise from the addition funnel at a rate sufficient to maintain reflux. During this time a white precipitate of NaCl forms. After the addition is complete, the reaction mixture is heated to maintain reflux for 10 hr, allowed to cool, and then dried on a rotary evaporator. The product is dissolved in 500 mL of hot acetone and the suspension filtered with suction on a medium frit to remove NaCl. (It may be necessary to wash the filter funnel with 0.5 L of hot acetone if the precipitate clogs it.) Slow cooling of the acetone solution gives white needles that are collected by filtration and dried *in vacuo* to yield the product (52.0 g, 84%). Alternatively, the acetone solution may be evaporated to dryness, dissolved as much as possible in hot chloroform (~0.5 L), filtered, and the solution allowed to crystallize. Thin layer chromatography (TLC) (silica gel, EtOAc): R_f = 0.42; ^{1}H NMR (300 MHz, δ, CDCl$_3$, TMS) 3.77 (t, 4H), 2.79 (m, 12H), 2.27 (s, 2H).

Anal. Calcd. for C$_8$H$_{18}$S$_3$O$_2$: C 39.64; H 7.48%. Found: C 39.69; H 8.06%.

1,1'-[Thiobis(2,1-ethanediylthio)]bis[2-chloroethane]. **Danger!** This and other 2-halo-ethyl sulfides (sulfur mustards) are powerful vesicants (blister-causing agents). The preparation must be carried out in an efficient fume hood, gloves (preferably two pairs) must be worn, and the utmost care must be taken to avoid contact with these compounds.

To a 500-mL round-bottomed flask with a magnetic stirring bar are added dry CH$_2$Cl$_2$ (200 mL; distilled from CaH$_2$) and 2,2'-[thiobis(2,1-ethanediylthio)]bis(ethanol) (8.0 g, 33.0 mmol). Freshly distilled thionyl chloride (8.0 mL, 0.11 mol) dissolved in 25 mL of dry CH$_2$Cl$_2$ is placed in a 50 mL, pressure-equalizing addition funnel that is connected to the flask by means of a double-necked adapter. The other arm of the adapter bears a Drierite® drying tube. Upon careful dropwise addition of the thionyl chloride to the stirred suspension, vigorous gas evolution occurs and the diol dissolves to give a colorless solution. After being stirred for 6 hr, the solution is treated with 5 mL of MeOH to quench excess SOCl$_2$ (■ **Caution:** *further gas evolution*) and is then rotary evaporated to dryness. Finally, it is pumped (0.5 mm) overnight to remove residual HCl. To minimize the risk of exposure to this hazardous compound, its yield is not determined but is assumed to be quantitative. TLC (CH$_2$Cl$_2$, silica gel) R_f 0.77.

Hexathia-18-crown-6. To the 1,1'-[thiobis(2,1-ethanediylthio)]bis(2-chloroethane) obtained in the previous step is added 150 mL of spectrograde

DMF from a freshly opened bottle (with gentle warming if necessary to dissolve the dichloride), followed by 2,2′-thiobis(ethanethiol) (5.09 g, 33.0 mmol). This solution is placed in a 250-mL Hirschberg constant addition funnel (Note 1) and is added under a nitrogen atmosphere over a 36-hr period to a suspension of cesium carbonate (Note 2) (13.0 g, 39.9 mmol) in 350 mL of DMF at 50–55° (Note 3). After the addition is complete, stirring at 50° is continued for an additional hour. The solvent is removed by vacuum distillation (at 1-mm pressure) and the residue is stirred with 300 mL of CH_2Cl_2 for 15 min, and the suspension is filtered through Celite®. The filtrate is washed three times with 80 mL of a 1 M aqueous NaOH solution, and once with 100 mL of H_2O, and then dried over anhydrous Na_2SO_4. The dried solution is evaporated *in vacuo* and the residue is recrystallized from hexane–acetone (4:1 v/v) by allowing a hot saturated solution to cool slowly to give the product. If necessary, the product can be purified further by flash chromatography[21] on silica gel with CH_2Cl_2 as eluant. Yield: 9.0 g (76%; 64% overall). An analogous procedure with appropriate modifications affords other crown thioethers in high yield.

Properties

Hexathia-18-crown-6 melts at 90–91°, yields a parent ion peak at m/e 360, and has R_f 0.35 on silica gel with CH_2Cl_2 as eluant. Its ^{1}H NMR in $CDCl_3$ shows a singlet at S = 2.82.

Anal. Calc. for $C_{12}H_{24}S_6$: C 39.96; H 6.71; S 53.34%. Found: C 40.04; H 6.60; S 53.35%.
    The IR 3400 (w,b), 2900 m, 1428 s, 1410 sh, 1310 (sh), 1269 (m), 1230 (w), 1202 (s), 1159 (m), 1130 (sh), 1030 (w,b), 962 (m), 878 (w,b), 842 (s), 738 (w), 709 (m), 694 (w), 676 (m). The ligand has also been structurally characterized.[5,6]

Note 1. Hirschberg funnels deliver solution through the gap between a tapered piston and cylindrical barrel, and include a pressure-equalizing arm. They are available from Kontes Scientific Glassware/Instruments, Spruce Street, P. O. Box 729, Vineland, NJ 08360, catalog No. K-634620.

Note 2. Finely milled potassium carbonate can be substituted for cesium carbonate with approximate halving of the yield.

Note 3. The addition funnel requires periodic checking as the DMF causes the valve to swell and thereby reduces the flow rate.

References

1. E. J. Hintsa, J. R. Hartman, and S. R. Cooper, *J. Am. Chem. Soc.*, **105**, 3738–8 (1983).
2. J. R. Hartman, E. J. Hintsa, and S. R. Cooper, *J. Chem. Soc. Chem. Commun.*, **1984**, 386–387.
3. J. R. Hartman, and S. R. Cooper, *J. Am. Chem. Soc.*, **108**, 1202–1208 (1986).
4. J. R. Hartman, E. J. Hintsa, and S. R. Cooper, *J. Am. Chem. Soc.*, **108**, 1208–1214 (1986).
5. J. R. Hartman, R. E. Wolf, Jr., B. R. Foxman, and S. R. Cooper, *J. Am. Chem. Soc.*, **105**, 131–132 (1983).
6. R. E. Wolf, Jr., J. R. Hartman, J. M. E. Storey, B. M. Foxman, and S. R. Cooper, *J. Am. Chem. Soc.* **109**, 4328–35 (1987).
7. W. N. Setzer, C. A. Ogle, G. S. Wilson, and R. S. Glass, *Inorg. Chem.*, **22**, 266–271 (1983).
8. S. C. Rawle, J. R. Hartman, D. J. Watkin, and S. R. Cooper, *J. Chem. Soc. Chem. Commun.*, **1986**, 1083–1084.
9. S. C. Rawle, and S. R. Cooper, *J. Chem. Soc. Chem. Commun.* **1987**, 308–9.
10. J. Clarkson, R. Yagbasan, P. J. Blower, S. C. Rawle, and S. R. Cooper, *J. Chem. Soc. Chem. Commun.* **1987**, 950–1.
11. S. C. Rawle, R. Yagbasan, K. Prout, and S. R. Cooper, *J. Am. Chem. Soc.* **109**, 6181–2 (1987).
12. P. W. R. Corfield, C. Ceccarelli, M. D. Glick, I. W.-Y. Moy, L. A. Ochrymowycz, and D. B. Rorabacher, *J. Am. Chem. Soc.*, **107**, 2399–2404 (1985).
13. L. L. Diaddario, Jr., E. R. Dockal, M. D. Glick, L. A. Ochrymowycz, and D. B. Rorabacher, *Inorg. Chem.*, **24**, 356–363 (1985).
14. V. B. Pett, L. L. Diaddario, Jr., E. R. Dockal, P. W. Corfield, C. Ceccarelli, M. D. Glick, L. A. Ochrymowycz, and D. B. Rorabacher, *Inorg. Chem.*, **22**, 3661–3670 (1983).
15. D. B. Rorabacher, M. J. Martin, M. J. Koenigbauer, M. Malik, R. R. Schroeder, J. F. Endicott, and L. A. Ochrymowycz, *Copper Coordination Chemistry: Biochemical and Inorganic Perspectives*, K. D. Karlin and J. Zubieta (eds.),Adenine Press, Guilderland NY, 1983 pp. 167–202 and references therein.
16. J. R. Meadow, and E. E. Reid, *J. Am. Chem. Soc.*, **56**, 2177–2180 (1934).
17. D. St. C. Black and I. A. McLean, *Tetrahedron Lett.*, **1969**, 3961–3964.
18. D. St. C. Black and I. A. McLean, *Aust. J. Chem.*, **24**, 1401–1411 (1971).
19. J. Buter, and R. M. Kellogg, *J. Chem. Soc. Chem. Commun.*, **1980**, 466–467.
20. L. A. Ochrymowycz, C.-P. Mak, and J. D. Michna, *J. Org. Chem.*, **39**, 2079–2084 (1974).
21. W. C. Still, M. Kahn, and A. Mitra, *J. Org. Chem.*, **43**, 2923–2925 (1978).

29. ([18]CROWN-6)POTASSIUM DICYANOPHOSPHIDE(1-)

Submitted by ALFRED SCHMIDPETER* and GÜNTHER BURGET*
Checked by DON J. CHANDLER† and RICHARD A. JONES†

$$4P_4 + 2MCN \longrightarrow MP(CN)_2 + MP_{15}$$

$$M^+ = (1,4,7,10,13,16\text{-hexaoxacyclooctadecane})\text{potassium}(1+)$$
$$= ([18]\text{crown-6})K(1+)$$

The dicyanophosphide(1-)* ion $P(CN)_2^-$ may be viewed as a homolog of the dicyanoamide(1-) ion $N(CN)_2^-$ and related to the thiocyanate ion SCN^- in the series of (pseudo)halide ions $E(CN)_{7-n}^-$ (n = group number of element E).[1] In this sense it can be used in reactions with various kinds of electrophiles to introduce the $P(CN)_2$ group.[1-6] It may also be viewed as a cyanide complex of cyanophosphinidene PCN or of the phosphorus(I) cation P^+, thus, strong nucleophiles replace one or both cyanide ions.[6-8]

The dicyanophosphide(1-) ion was first prepared by reduction of $P(CN)_3$.[1] It is more conveniently obtained from white phosphorus in a nucleophilic disproportionation by an ammonium, phosphonium, or crown ether alkali metal cyanide.[9]

Crystalline $[([18]crown-6)K][P(CN)_2]$ is the preferred material for further reactions. In the P_4-disproportionation reaction, crown ether potassium polyphosphides are formed as the second product. Their composition and solubility depend on the stoichiometry used. The intense red color of the solution in early stages of the reaction is caused by an intermediate high concentration of soluble polyphosphides. The final equilibrium concentration of these phosphides can contaminate and color the isolated dicyanophosphide. To avoid this contamination, the remaining soluble polyphosphides are converted to insoluble ones by an excess of white phosphorus.

■ **Caution.** *All the reagents used and the product are toxic. White phosphorus must be kept under water. It ignites on contact with air; the fumes generated are irritating and extremely poisonous. Contact of white phosphorus with the skin causes severe burns. Absorption in any form may be lethal. Even small amounts of potassium cyanide are lethal. For [18]-crown-6 (oral) an LD 50 of 705 mg/kg is reported. Crown ethers are skin and eye irritants. All used and emptied equipment is immediately rinsed with a dilute solution of bromine in methanol under a well-ventilated hood.*

Procedure

A dry 250-mL, two-necked, round-bottomed flask with an inert-gas inlet is equipped with a magnetic stirring bar and a reflux condenser topped by a paraffin oil filled pressure relief valve. The reaction is performed under dry and oxygen-free argon.† Traces of oxygen are removed from the inert

---

*Also known as dicyanophosphate.

†Oxygen-free nitrogen can be used instead but is less convenient.

---

*Institut für Anorganische Chemie, Universität München D-8000 München 2, Federal Republic of Germany.

†Department of Chemistry, The University of Texas at Austin, Austin, TX 78712.

gas by passing it over a chromium(II) contact.[10] Acetonitrile is dried by passing it through a column of 3-Å molecular sieve. To free them from water, bars of white phosphorus in a Schlenk tube are washed twice with dry THF, vacuum dried for 10 min and, without air contact, directly transferred into the reaction solution.

To a solution of 13.22 g (50 mmol) of [18]crown-6 [Aldrich] and 3.26 g (50 mmol) of potassium cyanide in 100 mL of dry acetonitrile, 12.4 g (100 mmol) of white phosphorus are added. The yellow solution is stirred and heated to reflux. While the phosphorus melts, the solution turns dark red. After boiling for some 2 hr, a black precipitate settles from the solution and after approximately 24 hr the red color of the solution fades somewhat. After 1 day at room temperature excess white phosphorus (3.0 g, 24 mmol) is added. During 24 hr of further boiling the solution becomes yellow and some phosphorus deposits in the condenser. The black precipitate is separated by filtering through a Schlenk-type glass frit and washed three times with 20 mL of acetonitrile. The solid residue, left after completely evaporating the filtrate, is washed three times with 50 mL of toluene and then vacuum dried, giving 7.60 to 8.00 g [([18]crown-6)K] [P(CN)$_2$] as a pale yellow to pink powder (79–83% of the theoretical yield based on KCN or crown ether).

Alternatively, the black precipitate formed after the first 24 hr is separated, and the excess phosphorus is added to the clear red filtrate. During the second 24 hr period of boiling some more of the black polyphosphide precipitate forms. [([18]crown-6)K] [P(CN)$_2$] is obtained in colorless and analytically pure form but in a somewhat reduced yield, 6.2 g (64%).

Anal. Calcd. for [C$_{12}$H$_{24}$O$_6$K]C$_2$N$_2$P (386.4): C, 43.51; H, 6.26; N, 7.25. Found: C, 43.26; H, 6.55; N, 7.25.

The yield based on the (expensive) crown ether may on the other hand be increased to 85% if a 20% excess of KCN is used (60 mmol instead of 50 mmol).

Properties

The salt [([18]crown-6)K] [P(CN)$_2$] melts at 130 to 133°. It is soluble in THF, acetonitrile, chloroform, dichloromethane, 1,2-dichloroethane, but insoluble in benzene, toluene, and diethyl ether. It is stable in boiling solvents such as acetonitrile or THF. It may be handled in air for a short time, but hydrolyzes with water or wet solvents. IR: 2113 (ν_sCN), 2090 (ν_{as}CN), 634 (ν_sPC), 620 cm^{-1} (ν_{as}PC). ^{31}P NMR (MeCN)[11]: δ = 193.9. ^{13}C NMR (MeCN): δ = 130.7, J_{PC} = 105.5 Hz.

References

1. A. Schmidpeter and F. Zwaschka, *Angew. Chem.*, **89**, 747 (1977); *Angew. Chem. Int. Ed. (Engl.)*, **16**, 704 (1977).
2. A. Schmidpeter and F. Zwaschka, *Angew. Chem.*, **91**, 441 (1979); *Angew. Chem. Int. Ed. (Engl.)*, **18**, 411 (1979).
3. A. Schmidpeter, W. Gebler, F. Zwaschka, and W. S. Sheldrick, *Angew. Chem.*, **92**, 767 (1980); *Angew. Chem. Int. Ed. (Engl.)*, **19**, 722 (1980).
4. A. Schmidpeter, F. Zwaschka, and W. S. Sheldrick, In *Phosphorus Chemistry: Proceedings of the 1981 International Conference* (ACS Symposium Series 171) American Chemical Society, Washington, DC, 1981, p. 419.
5. A. Schmidpeter and F. Zwaschka, *Z. Chem.*, **24**, 376 (1984).
6. A. Schmidpeter and G. Burget, *Z. Naturforsch.*, B **40**, 1306 (1985).
7. A. Schmidpeter, S. Lochschmidt, G. Burget, and W. S. Sheldrick, *Phosphorus Sulfur*, **18**, 23 (1983).
8. A. Schmidpeter, K.-H. Zirzow, G. Burget, G. Huttner, and I. Jibril, *Chem. Ber.*, **117**, 1695 (1984).
9. A. Schmidpeter, G. Burget, F. Zwaschka, and W. S. Sheldrick, *Z. Anorg. Allg. Chem.*, **527**, 17 (1985).
10. Reduced Phillips catalyst: H. L. Krauss and H. Stach, *Z. Anorg. Allg. Chem.*, **366**, 34 (1969).
11. S. Lochschmidt and A. Schmidpeter, *Phosphorus Sulfur*, **29**, 73 (1987).

30.   (2-DIPHENYLPHOSPHINO)BENZENAMINE

Submitted by MERVYN K. COOPER,* J. MICHAEL DOWNES,*
and PAUL A. DUCKWORTH†
Checked by MICHAEL C. KERBY,‡ RONALD J. POWELL,‡
and MARK D. SOUCEK‡

The preparation of tertiary phosphine ligands is most commonly achieved by standard addition reactions between alkali metal phosphides and organohalides, or between chlorophosphines and organolithium or Grignard reagents.[1] Such syntheses, however, are often unsuitable for 2-substituted phenylphosphines due to the reactivity or steric requirements of the non-phosphine functionality. This was the case with the bidentate ligand (2-diphenylphosphino)benzenamine, H_2L, where the previously reported synthesis[2] was based on the high temperature reduction of the phosphine oxide with polymethylhydrosiloxane. The method of preparation described

*School of Chemistry, University of Sydney, Sydney N.S.W. 2006, Australia.
†Now at the Research School of Chemistry, The Australian National University, Canberra A.C.T. 2601, Australia.
‡Department of Chemistry, University of Texas, Austin, TX 78712-1167.

here makes use of low-cost starting materials, is less time consuming (4–5 days as opposed to up to 10 days for the original method) and is suitable, with only minor modification,[3] for the synthesis of (2-diphenylphosphino)-*N*-methylbenzenamine and (2-diphenylphosphino)phenol.[4,5]

The title ligand, H_2L, forms complexes with the later transition metals, which can be deprotonated[2,6] to produce compounds containing the little studied phenylamido donor group. Its complexes with platinum and rhodium have been shown to be of mechanistic significance.[7,8]

$$\text{(1)}$$

$$\text{(2)}$$

$$\text{(3)}$$

■ **Caution.** *The vapor of 2-chlorobenzenamine is highly toxic, the liquid can also be absorbed through the skin. In addition, both anhydrous nickel chloride and dry (2-aminophenyl)triphenylphosphonium chloride tend to produce a fine dust when handled. It is recommended that procedures involving these materials be conducted in an efficient hood.*

Procedure for Reactions 1 and 2

A round-bottomed, 500-mL flask equipped with a Teflon-coated magnetic stirring bar, still-head and condenser is charged with 2-chlorobenzenamine, 128 g (1.00 mol), and triphenylphosphine, 262 g (1.00 mol). Next, 65 g (0.50 mol) of powdered, anhydrous nickel chloride (dried by heating the hexahydrate at 130° for at least 48 hr) is added and the mixture is heated to 200° with stirring. A small amount of residual water distills over. The temperature is maintained at 200–220° for 4 hr. The resulting dark blue melt is cooled to 180–160° then poured **(CARE!)** into 600 mL of hot (60°) water previously acidified with a few drops of concentrated HCl. Boiling acidified water (400 mL) is added to the reaction vessel to extract any remaining material. The combined extracts are stirred while hot until the blue color of the melt is completely discharged. After cooling and washing with diethyl ether (1 × 300, 2 × 200 mL), to remove unreacted starting materials, the aqueous phases are extracted with dichloromethane (3 × 300 mL). The combined dichloromethane extracts are dried over anhydrous Na_2SO_4 and evaporated to an orange oil (about 300 mL). (If all of the dichloromethane is evaporated or if the mixture is allowed to cool after evaporation then the resulting oil may be so viscous as to prevent efficient stirring in the next step.) Tetrahydrofuran (about 500 mL) is added with vigorous stirring until white crystals begin to form. The mixture is stored at about 5° for several hours. The white crystalline mass is collected by filtration, washed with THF then diethyl ether and dried (120°, about 15 torr) overnight to remove solvent of crystallization to yield 222 g (57%) of the analytically pure product as a white powder, mp 293–295°.

Anal. Calcd. for $C_{24}H_{21}ClNP$: C, 73.9; H, 5.4; Cl, 9.1; N, 3.6; P, 8.0. Found: C, 73.6; H, 5.5; Cl, 9.3; N, 3.6; P, 8.0.

Procedure for Reaction 3

An oven dried, three-necked, round-bottomed 1-L flask is charged with 500 mL of anhydrous THF (freshly distilled from sodium benzophenone ketyl), 51 g (0.39 mol) of naphthalene and a Teflon-coated, magnetic stirring bar. Sodium wire, 8.3 g (0.36 mol) is extruded into the vessel, which is then quickly fitted with a condenser and nitrogen inlet. The mixture is stirred in a dry nitrogen atmosphere until the sodium has completely dissolved (about 1 hr). The resulting very dark green solution is cooled in a solid CO_2–acetone bath until it is almost completely solid (inadequate cooling of the naphthalenide solution leads to reduced yields in the subsequent reduction). (2-Aminophenyl)triphenylphosphonium chloride, 64.0 g (0.164 mol), obtained in the previous reaction, is added and the mixture allowed to warm slowly, with occasional shaking, to room temperature, then stirred for 1 hr. Acetic acid, about 2 g (0.03 mol), is added dropwise to discharge the last of the green color. The orange-red mixture is treated slowly with 100 mL of a 20% ammonium chloride solution and sufficient water is added to dissolve any remaining solid (no further precautions to exclude atmospheric oxygen need be taken from this point). The two resulting phases are separated and the aqueous layer is extracted with diethyl ether (1 × 100 mL). The combined organic phases are dried over anhydrous Na_2SO_4 and evaporated. The residue is taken up in 500 mL of boiling 90% ethanol and treated with a solution of nickel nitrate hexahydrate, 26 g (0.09

mol) in 100 mL of boiling 90% ethanol. The resulting brown solution is treated with 2 mL of trifluoroacetic acid to ensure that deprotonation of the amino group does not occur. After storing overnight at 5° the mixture of metal complex and crystallized naphthalene is filtered off and washed first with ethanol and then with diethyl ether to remove the naphthalene. A yield of 55.4 g (89%) of orange crystals, $[Ni(H_2L)_2](NO_3)_2 \cdot H_2O^2$, is obtained.

The nickel complex obtained is suspended in a mixture of benzene (300 mL) and water (300 mL) to which a few drops of concentrated HCl are added. The mixture is refluxed until the crystals dissolve (several hours). The two resulting phases are separated and the aqueous layer is extracted with benzene (50 mL). The combined organic extracts are washed with brine (100 mL), dried over anhydrous Na_2SO_4 then passed down a short alumina column (2.5 × 10 cm) eluting with benzene. The now colorless solution is evaporated to an oil, which is taken up in 180 mL of boiling 90% ethanol. Fluffy white crystals of the analytically pure ligand, mp 82 to 83°, form on cooling. The yield is 35.6 g (88% from the nickel complex). The mother liquor may be evaporated, the residue taken up in boiling 90% ethanol and treated with nickel nitrate solution as above to give 2.5 g (5%) of recovered nickel complex.

Anal. Calcd. for $C_{18}H_{16}NP$: C, 78.0; H, 5.8; N, 5.1; P, 11.2. Found: C, 78.1; H, 5.8; N, 5.0; P, 11.0.

Properties

The title ligand, H_2L, is an air-stable, crystalline solid soluble in most organic solvents. Both it and its phenylphosphonium salt are slightly light sensitive and should be stored in dark containers. The IR spectrum (Nujol mull) exhibits bands at 3455 and 3365 cm^{-1} attributable to ν_{N-H}. The phosphorus nucleus resonates at $\delta_{CH_2Cl_2} = -21.0$ (relative to 85% H_3PO_4) in the ^{31}P {^1H} NMR spectrum. Treatment of H_2L with benzoyl chloride gives the *N*-benzoyl derivative, the iridium(I) complex of which undergoes oxidative addition across the N—H bond.[9,10]

References

1. C. A. McAuliffe and W. Levason, *Studies in Inorganic Chemistry 1: Phosphine, Arsine and Stibine Complexes of the Transition Elements*, Elsevier, Amsterdam, 1979.
2. M. K. Cooper and J. M. Downes, *Inorg. Chem.*, **17**, 880 (1978).

3. M. K. Cooper, J. M. Downes, and P.A. Duckworth, *Aust. J. Chem.*, submitted for publication.
4. H. D. Empsall, B. L. Shaw, and B. L. Turtle, *J. Chem. Soc. Dalton Trans.*, **1976**, 1500.
5. T. B. Rauchfauss, *Inorg. Chem.*, **16**, 2966 (1977).
6. C. W. G. Ansell, M. McPartlin, P. A. Tasker, M. K. Cooper, and P. A. Duckworth, *Inorg. Chim. Acta*, **76**, L135 (1983).
7. M. K. Cooper and J. M. Downes, *J. Chem. Soc. Chem. Commun.*, **1981**, 381.
8. G. J. Organ, M. K. Cooper, K. Henrick, and M. McPartlin, *J. Chem. Soc. Dalton Trans.*, **1984**, 2377.
9. D. Hedden, D. M. Roundhill, W. C. Fultz, and A. L. Rheingold, *J. Am. Chem. Soc.*, **106**, 5014 (1984).
10. D. Hedden and D. M. Roundhill, *Inorg. Chem.*, **25**, 9 (1986).

31. SODIUM SALT OF (1*R*)-3-NITROBORNAN-2-ONE (SODIUM *d*-α-CAMPHORNITRONATE)*

$$C_{10}H_{15}OBr + HNO_3 \longrightarrow C_{10}H_{14}BrNO_3 + H_2O$$

$$2Na + 2C_2H_5OH \longrightarrow 2NaOC_2H_5 + H_2 \uparrow$$

$$C_{10}H_{14}BrNO_3 + 2NaOC_2H_5 \longrightarrow$$
$$NaC_{10}H_{14}NO_3 + NaBr + C_2H_5OH + CH_3CHO$$

The last equation is not certain.

Submitted by JONATHAN W. STOCKER† and JOHN C. BAILAR, JR.†
Checked by GEORGE B. KAUFFMAN,‡ PHILIP CHU,‡
and RONALD L. MARHENKE‡

The sodium salt of (1*R*)-3-nitrobornan-2-one (sodium-*d*-α-camphornitronate) has been used as a resolving agent for inorganic complex cations on several occasions.[1] It is best prepared by a two-step synthesis. First, crude (1*R*)-3-*endo*-bromo-3-*exo*-nitrobornan-2-one (*d*-α,α-nitrobromocamphor) is prepared by reaction of (1*R*)-3-*endo*-bromobornan-2-one (*d*-bromocamphor) with concentrated nitric acid. The crude *d*-α,α-nitrobromocamphor is then treated with sodium ethoxide to form $Na(d\text{-}C_{10}H_{14}NO_3)$. The following procedure is a modification of the methods given by Lowry and Steele[2,3] and by Clapp.[4]

*(+)-1,7,7-Trimethyl-3-aci-nitrobicyclo[2.2.1]-2-one sodium salt.
†Department of Chemistry, University of Illinois, Urbana, IL 61801.
‡Department of Chemistry, California State University, Fresno, CA 93740.

Procedure

Mix 150 g of (1*R*)-3-*endo*-bromobornan-2-one (*d*-bromocamphor) (0.65 mol) with 500 mL of concentrated (15.8 *M*) nitric acid (7.90 mol) in a 1-L round-bottomed flask connected by ground glass connections to two long, water-cooled condensers in order to prevent escape of the product. (■ **Caution.** *The reaction must be carried out in a well-ventilated hood because toxic nitrogen dioxide is produced as a by-product.*) The mixture is heated carefully with magnetic stirring in an oil bath on a hot plate until its temperature reaches 118°, at which constant temperature the mixture boils. (The reaction starts at about 85°, as evidenced by evolution of orange-brown nitrogen dioxide gas.) Once the reaction has started, no further heating is necessary for about 20 min. After the initial reaction subsides, the oil bath is heated to about 125° to keep the mixture at 114°, just below its boiling point (about 140°). After 2.5 hr of heating, the mixture has become a translucent pale yellow. The temperature is maintained for 40 hr, after which the mixture is allowed to cool, whereupon it becomes viscous and partially solidifies. The supernatent liquid is removed with a pipet, and the residue is poured into 200 mL of water, where it solidifies. The supernatant is decanted. The air-dried residue consists of about 70 g of crude (1*R*)-3-*endo*-bromo-3-*exo*-nitrobornan-2-one (*d*-α,α-nitrobromocamphor) (39.06% yield).

The crude *d*-α,α-nitrobromocamphor is dissolved with heating in 40 mL of *absolute* ethanol. Place 15.5 g (0.674 g-atom) of sodium, dissolved in 135 mL of absolute alcohol, into a 500-mL round-bottomed flask that has been fitted with a long reflux condenser. (■ **Caution.** *This vigorous, exothermic reaction, which results in the evolution of hydrogen, should be carried out in a well-ventilated hood.*) The dissolved (*d*-α,α-nitrobromocamphor is poured very slowly into the 500-mL flask containing the sodium ethoxide. During the addition, continuous stirring with a glass rod is required. (If the *d*-α,α-nitrobromocamphor solution is poured only slowly down the condenser it solidifies in the condenser. The checkers suggest that the condenser may be removed to facilitate the addition.) The flask is cooled in ice during this addition, resulting in almost immediate solidification. The precipitated sodium salt of (1*R*)-3-nitrobornan-2-one (sodium *d*-α-camphornitronate) is collected by filtration. The yield is 60.40 g (42.45%). The mother liquor can be frozen partially in a dry ice–acetone bath and more of the sodium salt removed by quickly filtering the mother liquor before the salt has had time to redissolve. In this way an additional 2.11 g (1.48%) of the product can be obtained. The pale yellow crystals are washed once with a minimum volume (40 mL) of a 1:1 mixture of acetone and *absolute* ethanol. They are recrystallized from a minimum volume (540 mL) of *absolute* ethanol and air dried. The yield is about 43.76 g (30.75%). If desired, the product may be dried further over phosphorus(V) oxide at 80°, whereupon it suffers a weight loss of 11.25%, resulting in a final yield of 38.84 g (27.29%). Its specific rotation $[\alpha]_D^{20} = +285°$.

Anal. Calcd. for $NaC_{10}H_{14}NO_3$: C, 54.79; N, 6.39, H, 6.44. Found: C, 53.5; N, 6.1; H, 6.4.*

Properties

The sodium salt of $(1R)$-3-nitrobornan-2-one (sodium *d*-camphornitronate) is a pale yellow crystalline solid, easily soluble in water. It is slightly soluble in absolute ethanol and almost completely insoluble in acetone. Its optical activity has been measured;[5] $[\alpha]_D^{20} = +295°$; $[\alpha]_{5750}^{20} = +308°$; $[\alpha]_{5461}^{20} = +368°$.

Acknowledgments

The submitters wish to thank the checkers for rewriting the synthesis, which has resulted in a great improvement.

References

1. See, for example, A. Werner, *Ber. Chem. Ges.* **45**, 865 (1912).
2. T. M. Lowry, *J. Chem. Soc.,* **73**, 995 (1898).
3. T. M. Lowry and V. Steele, *J. Chem. Soc.,* **107**, 1039 (1915).
4. L. B. Clapp, *The Stereochemistry of Complex Inorganic Compounds,* Ph.D. dissertation, University of Illinois, Urbana, IL, 1941, p. 13.
5. H. D. K. Drew, F. S. H. Head, and H. J. Tress, *J. Chem. Soc.,* **1937**, 1549.

32. TRIS(GLYCINATO)COBALT(III)

$$10KHCO_3 + 2CoCl_2 \cdot 6H_2O + H_2O_2 \longrightarrow$$
$$2K_3[Co(CO_3)_3] + 4CO_2 \uparrow + 4KCl + 18H_2O$$

$$K_3[Co(CO_3)_3] + 3H_2NCH_2COOH + 3HC_2H_3O_2 \longrightarrow$$
$$fac\text{-} \text{ and } mer\text{-}[Co(H_2NCH_2COO)_3] + 3KC_2H_3O_2 + 3CO_2 \uparrow + 3H_2O$$

---

*The checkers obtained the following values: C, 52.23; H, 6.08; N, 6.02. The fact that each of these values is approximately 5% low is believed to be due to the presence of sodium bromide in the product.

**Submitted by GEORGE B. KAUFFMAN,* MOHAMMAD KARBASSI,* and
EISHIN KYUNO†**
Checked by W. J. BIRDSALL‡ and P. E. A. KYLANPAA‡

One of the earliest cases of stereoisomerism among inner complexes, tris(glycinato)cobalt(III),[1] occurs in two forms, violet (α) and red (β). In one of these geometric isomers, all of the amino groups of the glycine molecules are adjacent (*fac*), while in the other, two of these occupy opposite positions (*mer*). Absorption spectra indicate that the more soluble violet α form is the *mer* isomer.[2–5]

Three general preparative methods are available:[6]

1. The classic method of Ley and Winkler[1]—dissolving freshly prepared cobalt(III) hydroxide oxide, $CoO(OH)$, in an aqueous solution of glycine with heating—gives both products, with the *mer* isomer predominating (often this is the only isomer formed).
2. The method of Neville and Gorin[7]—allowing glycine to react with hexaamminecobalt(III) chloride in a boiling aqueous potassium hydroxide solution under reflux—favors formation of the *fac* isomer.
3. The method of Shibata et al.,[8] employed here—allowing glycine to react with aqueous potassium tri(carbonato)cobaltate(III)—gives both isomers in approximately equal amounts.

All three methods have been applied to synthesize tris complexes of other amino acids,[8,9] including those with optically active acids.[10–16]

Procedure

Thirty grams of potassium hydrogencarbonate (0.300 mol) is added to 30 mL of water contained in a 150-mL beaker, and the mixture (**A**) is cooled in an ice bath for 15 min with mechanical stirring. In a separate 100-mL beaker, 10.0 g of cobalt(II) chloride 6-hydrate (0.042 mol) is added to 10 mL of water at 30°. The resulting mixture is allowed to stand for 15 min in an ice bath, and 15 mL of 30% hydrogen peroxide is then added slowly to it (solution **B**). Solution **B** is added to **A** at a rate of one drop every 5 sec, with mechanical stirring at 0–5°, followed by suction filtration. Nine grams of glycine (0.120 mol) is added to the resulting green filtrate of potassium tri(carbonato)cobaltate(III). Attempts to prepare tris-

*Department of Chemistry, California State University, Fresno, CA 93740.
†School of Pharmacy, Hokuriku University, 3, Ho Kanagawa-machi, Kanazawa 920-11, Japan.
‡Albright College, P.O. Box 516, Reading, PA 19603.

(glycinato)cobalt(III) from sodium tri(carbonato)cobaltate(III) 3-hydrate[17] resulted in poor yields and little of the *fac* isomer.

The mixture is heated at 60–70° on a water bath until the color of the resulting solution changes from green to dark blue to violet (~30 min). Then 21 mL of 6 *N* acetic acid is added slowly at a rate of 1 drop every 5 sec with mechanical stirring at 60–70° (effervescence). A slight excess (0.5 mL) of acetic acid may be added to ensure completion of the reaction. The solution is stirred vigorously until the evolution of carbon dioxide ceases and the color of the solution has become reddish violet.

The solution is concentrated to two thirds of its original volume and is allowed to stand overnight. The deposited reddish pink crystals of the less soluble *fac*(β) isomer are collected by suction filtration on a 60-mL sintered glass funnel (medium porosity), washed successively with three 10-mL portions each of cold water, ethanol, and diethyl ether, and dried in a vacuum desiccator over Drierite®, followed by 2 hr of additional drying at 100° *in vacuo*. The filtrate is concentrated in a rotary evaporator until violet crystals of the more soluble *mer*(α) isomer are deposited. These are collected on a 60-mL sintered glass funnel (medium porosity) and washed and dried in the same manner as was the *fac*(β) isomer. The yields of *fac* and *mer* isomers are 2.1 g (18.7%) and 2.7 g (24.0%), respectively. Conditions have been chosen to yield appreciable amounts of both isomers. Slight variations in the procedure result in different ratios of isomers. For example, if, before addition of the acetic acid, the reaction mixture is heated for an hour after it has become violet, the *fac* isomer is favored: *fac,* 3.3 g (29.3%); *mer,* 1.5 g (13.3%). On the other hand, if it is heated only until it has become dark blue (about 15 min), the *mer* isomer is favored: *fac,* 0.9 g (8.0%); *mer,* 4.2 g (37.4%).

Anal. fac Isomer Calcd. for $CoC_6H_{12}N_3O_6$: C, 25.64; H, 4.30; N, 14.95. Found: C, 25.61; H, 4.29; N, 14.75. *mer* Isomer Calcd. for $CoC_6H_{14}N_3O_7$: C, 24.09; H, 4.72; N, 15.05. Found: C, 24.03; H, 4.85; N, 14.15.

Visible spectrum (H_2O): 520, 372 nm (*fac*); 540, 370 nm (*mer*).

The procedure can be modified to prepare the corresponding complexes of other amino acids. For example, if 11.0 g of alanine is substituted for the glycine and if, before addition of the acetic acid, the reaction mixture is heated for an hour after it has become violet, the yields of *fac* and *mer* isomers of $[Co(H_2NCH_2CH_2COO)_3]$ are 1.5 g (11.6%) and 3.7 g (28.6%), respectively.

Anal. Calcd. for $CoC_9H_{18}N_3O_6$: C, 33.45; H, 5.61; N, 13.00. Found: *fac* Isomer, C, 33.39; H, 5.43; N, 12.97; *mer* Isomer, C, 33.42; H, 5.58; N, 12.85.

Visible spectrum (H_2O): 514, 366 (*fac*); 538, 369 nm (*mer*).

Properties

According to Ley and Winkler,[1] *mer*(α)-tris(glycinato)cobalt(III) forms large, dark violet, rhombic crystals containing two molecules of water of crystallization, while *fac*(β)-tris(glycinato)cobalt(III) forms reddish pink, needlelike crystals with one molecule of water of crystallization. Contrary to these results, our analytical and thermogravimetric data have shown the pink form to be anhydrous and the violet form to be a monohydrate. The number of molecules of water of crystallization varies with conditions of drying.

The isomers are not directly convertible. They are soluble only with difficulty in water (at 25°, 0.199 g red/L and 9.33 g violet/L;[1] at 20°, 0.192 g red/L and 9.21 g violet/L[18]) and in basic solvents such as pyridine and aniline. Both are considerably more soluble in acid solutions, from which they can be precipitated unchanged by addition of ethanol. As further evidence of their extraordinary stabilities, they can be recovered unchanged from hot concentrated sulfuric acid and can be treated with hot concentrated nitric acid for some time without noticeable decomposition. Both are decomposed, however, by prolonged warming with concentrated hydrochloric acid, resulting in evolution of chlorine and formation of cobalt(II) chloride. They are virtually insoluble in most organic solvents. Electrical conductance measurements of aqueous and sulfuric acid solutions of both isomers show them to be nonelectrolytes, while cryoscopic studies show them to be monomeric and undissociated in solution. Being unsymmetrical, both isomers can exist in enantiomorphic forms, but for many years no attempts at resolution were made, probably because their nonelectrolytic character is not suited to salt formation. By chromatographic adsorption on a starch column Krebs and Rasche[19] have resolved the violet isomer into its optical antipodes.

Neither isomer reacts with ammonia at low temperature. Both react with potassium cyanide giving potassium hexacyanocobaltate(III) and potassium glycinate. The red isomer reacts readily with potassium nitrite to give a solution yielding ruby red crystals, whereas the violet isomer undergoes this reaction only with difficulty.

References

1. H. Ley and H. Winkler, *Berichte der Deutschen Chemischen Gesellschaft*, **42,** 3894 (1909); **45,** 372 (1912).
2. H. Kuroya and R. Tsuchida, *Bull. Chem. Soc. Jpn.*, **15,** 427 (1940); Y. Shimura and R. Tsuchida, *Bull. Chem. Soc. Jpn.*, **29,** 311 (1956); K. Nakamoto, J. Fujita, M. Kobayashi, and R. Tsuchida, *J. Chem. Phys.*, **27,** 439 (1957).
3. F. Basolo, C. J. Ballhausen, and J. Bjerrum, *Acta Chem. Scand.*, **9,** 810 (1955).

4. A. J. Saraceno, I. Nakagawa, S. Muzishima, C. Curran, and J. V. Quagliano, *J. Am. Chem. Soc.*, **80**, 5018 (1958).
5. E. Fluck (ed.), *Gmelins Handbuch der Anorganischen Chemie*, 8th ed., Kobalt, System Nr. 58, Teil B, Ergänzungsband, Lieferung 2, Springer-Verlag, Berlin, 1964, p. 647.
6. M. Shibata, *Modern Syntheses of Cobalt(III) Complexes*, Springer-Verlag, Berlin, 1983, pp. 34–35.
7. R. G. Neville and G. Gorin, *J. Am. Chem. Soc.*, **78**, 4895 (1956).
8. M. Mori, M. Shibata, E. Kyuno, and M. Kanaya, *Bull. Chem. Soc. Jpn.*, **34**, 1837 (1961).
9. M. B. Ćelap, T. J. Niketić, T. J. Nibolić, and V. N. Nibolić, *Inorg. Chem.*, **6**, 2063 (1967).
10. I. Lifschitz, *Proc. K. Ned. Akad. Wet.*, **15**, 721 (1924); I. Lifschitz and W. Froentjes, *Rec. Trav. Chim.*, **60**, 225 (1941).
11. B. E. Douglas and S. Yamada, *Inorg. Chem.*, **4**, 1651 (1965).
12. J. H. Dunlop and R. D. Gillard, *J. Chem. Soc.* (*A*), **1965**, 6531; R. D. Gillard and N. C. Payne, *J. Chem. Soc.* (*A*), **1969**, 1197; R. D. Gillard, S. H. Laurie, D. C. Price, D. A. Phipps, and C. F. Weick, *J. Chem. Soc. Dalton Trans.*, **1974**, 1385; M. G. B. Drew, J. H. Dunlop, R. D. Gillard, and D. Rogers, *Chem. Commun.*, **1966**, 42.
13. T. Yasui, J. Hidaka, and Y. Shimura, *Bull. Chem. Soc. Jpn.*, **38**, 2025 (1965).
14. E. Larsen and S. F. Mason, *J. Chem. Soc.* (*A*), **1966**, 313.
15. R. G. Denning and T. S. Piper, *Inorg. Chem.*, **5**, 1056 (1966).
16. M. Shibata, H. Nishikawa, and K. Hosaka, *Bull. Chem. Soc. Jpn.*, **40**, 236 (1967); K. Hosaka, H. Nishikawa, and M. Shibata, *Bull. Chem. Soc. Jpn.*, **42**, 277 (1969).
17. H. F. Bauer and W. C. Drinkard, *Inorg. Synth.*, **8**, 202 (1966).
18. H.-G. Rosenkranz, Ph.D. Dissertation, Universität Jena, 1950, pp. 7, 11, 20.
19. H. Krebs and R. Rasche, *Z. Anorg. Chem.*, **276**, 236 (1954).

33. RESOLUTION OF THE TRIS(OXALATO)CHROMATE(III) ION BY A SECOND-ORDER ASYMMETRIC SYNTHESIS

Submitted by GEORGE B. KAUFFMAN,* NOBUYUKI SUGISAKA,† and IAN K. REID‡
Checked by R. KENT MURMANN§

The tris(oxalato)chromate(III) ion, $[Cr(C_2O_4)_3]^{3-}$, possesses the double historical distinction of being both the first resolved complex anion and the first resolved complex that did not contain nitrogen.[1] Since it racemizes rapidly in aqueous solution, more so than the corresponding cobalt(III) ion,[2,3] virtually all of this labile complex can be separated as a single en-

*Department of Chemistry, California State University, Fresno, CA 93740.
†Riker Laboratories, Inc., 270-4S-02 3M Center, St. Paul, MN 55144.
‡Research School of Chemistry, The Australian National University, Canberra 2600, Australia.
§Department of Chemistry, University of Missouri, Columbia, MO 65211.

antiomer, which precipitates as the less soluble diastereomer. In contrast to most resolutions, which employ low temperatures to keep racemization to a minimum, the technique of second-order asymmetric transformation[1,4] involves warming the solution during diastereomer formation in order to induce racemization.

The resolving agent, (+)-bis(1,2-ethanediamine)(oxalato)cobalt(III) iodide, is first converted to the acetate, which is then warmed with a solution of racemic potassium tris(oxalato)chromate(III) in order to precipitate the (+),(+) diastereomer by a second-order asymmetric transformation. The diastereomer is decomposed with potassium iodide, which also regenerates the resolving agent. The (+) enantiomer, which racemizes rapidly, is *immediately* precipitated from the filtrate with ethanol. An analogous procedure employing the (−) form of the resolving agent can be used to obtain the (−) enantiomer. The specific rotations obtained by this procedure ($[\alpha]_D = \pm1900°$) are higher than those reported previously by Jaeger[5] ($\pm420°$), Johnson and Mead[6] ($\pm1170°$), and Dwyer and Sargeson and co-workers[7] ($\pm1640°$).

A. PRECIPITATION OF (+)-BIS(1,2-ETHANEDIAMINE)-(OXALATO)COBALT(III) (+)-TRIS(OXALATO)-CHROMATE(III) HEXAHYDRATE, (+),(+) DIASTEREOMER

$$(+)\text{-}[Co(C_2O_4)(en)_2]I + AgC_2H_3O_2 \longrightarrow$$

$$(+)\text{-}[Co(C_2O_4)(en)_2]C_2H_3O_2 + AgI \downarrow$$

$$3\{(+)\text{-}[Co(C_2O_4)(en)_2]C_2H_3O_2\}$$

$$+ K_3(\pm)\text{-}[Cr(C_2O_4)_3]\cdot3H_2O + 3H_2O \longrightarrow$$

$$(+)\text{-}[Co(C_2O_4)(en)_2]_3(+)\text{-}[Cr(C_2O_4)_3]\cdot6H_2O \downarrow + 3KC_2H_3O_2$$

(en = 1,2-ethanediamine)

Procedure

Five grams (0.0127 mol) of (+)-bis(1,2-ethanediamine)(oxalato)-cobalt(III) iodide[8] ($[\alpha]_D^{25} = +720°$) is converted to the acetate by shaking for 10 min with 2.11 g (0.0126 mol) of solver acetate suspended in 30 mL of hot (50°) water contained in a 100-mL flask. The precipitated silver iodide is removed by filtration and washed with 10 mL of hot water. The combined red filtrate and washings are added with stirring to a solution of 2.00 g (0.00409 mol) of racemic potassium tris(oxalato)chromate(III) trihydrate[9,10] in 10 mL of water contained in a 100-mL beaker. The red-brown diastereomer begins to precipitate almost immediately. The suspension is quickly warmed to 50°, maintained at that temperature for 5 min, and allowed to cool slowly to 35°. The diastereomer crystals are collected by filtration on a 5-cm Büchner funnel, washed successively with 5-mL portions of ice water, 60% aqueous ethanol, 95% ethanol, and acetone, and are then air dried. The yield is about 3.7 g [74%, assuming complete racemization of the (−) enantiomer to the (+) enantiomer]. A 0.02% aqueous solution in a 1-dm tube gives $\alpha_D^{25} = +0.27°$, from which $[\alpha]_D^{25} = +1300°$.

Anal. Calcd. for $[Co(C_2O_4)(C_2H_8N_2)_2]_3[Cr(C_2O_4)_3] \cdot 6H_2O$: C, 23.52; H, 4.94; N, 13.72. Found: C, 23.70; H, 4.99; N, 13.74.

Smaller second (0.4 g) and third (0.1 g) fractions of diastereomer may be obtained by allowing the filtrate to stand at room temperature for 24 and 48 hr, respectively. Since the diastereomer racemizes only slowly in solution, all three fractions may be collected and combined (~4.2 g) (~84%) before proceeding with the following section.

B. ISOLATION OF POTASSIUM (+)-TRIS(OXALATO)-CHROMATE(III) DIHYDRATE

$$(+)-[Co(C_2O_4)(en)_2]_3(+)-[Cr(C_2O_4)_3] \cdot 6H_2O + 3KI \longrightarrow$$
$$K_3(+)-[Cr(C_2O_4)_3] \cdot 2H_2O + 3\{(+)-[Co(C_2O_4)(en)_2]I\}\downarrow + 4H_2O$$

Procedure

Since the optical enantiomers racemize rapidly in solution, all operations should be carried out as quickly as possible using iced solutions and iced apparatus.

The (+),(+) diastereomer (~4.2 g) is suspended in 15 mL of water contained in a 30-mL beaker, and 4 mL of saturated potassium iodide solution is added gradually with constant stirring. The precipitated resolving agent ($[\alpha]_D^{25} = +720°$) is recovered (in yields as high as 3.5–4.0 g or 70–80%) by filtration and is washed with a few milliliters of potassium iodide solution. Potassium (+)-tris(oxalato)chromate(III) dihydrate is *immediately* precipitated as bluish-mauve crystals from the combined filtrate and washings by slow addition of ice-cold ethanol until precipitation appears almost complete (10–20 mL). The product is collected by filtration on a 5-cm Büchner funnel, washed successively with 5-mL portions of ice-cold 80% aqueous ethanol, 95% ethanol, acetone, and is then air dried.

The yield is 1.43 g [75%, based on $K_3(\pm)$-$[Cr(C_2O_4)_3]\cdot 3H_2O$ or 89%, based on the $(+),(+)$ diastereomer]. A 0.02% aqueous solution in a 1-dm tube gives $\alpha_D^5 = +0.38°$, from which $[\alpha]_D^5 = +1900°$. Recrystallization does not increase the rotation.

Anal. Calcd. for $K_3[Cr(C_2O_4)_3]\cdot 2H_2O$: C, 15.35; H, 0.86. Found: C, 15.54; H, 0.89.

C. ISOLATION OF POTASSIUM ($-$)-TRIS(OXALATO)-CHROMATE(III) MONOHYDRATE

$(-)$-$[Co(C_2O_4)(en)_2]I + AgC_2H_3O_2 \longrightarrow$
$$(-)\text{-}[Co(C_2O_4)(en)_2]C_2H_3O_2 + AgI \downarrow$$

$3\{(-)\text{-}[Co(C_2O_4)(en)_2]C_2H_3O_2\}$
$$+ K_3(\pm)\text{-}[Cr(C_2O_4)_3]\cdot 3H_2O + 3H_2O \longrightarrow$$
$$(-)\text{-}[Co(C_2O_4)(en)_2]_3(-)\text{-}[Cr(C_2O_4)_3]\cdot 6H_2O \downarrow + 3KC_2H_3O_2$$

$(-)$-$[Co(C_2O_4)(en)_2]_3(-)$-$[Cr(C_2O_4)_3]\cdot 6H_2O + 2KI \longrightarrow$
$$K_3(-)\text{-}[Cr(C_2O_4)_3]\cdot H_2O + 3\{(-)\text{-}[Co(C_2O_4)(en)_2]I\}\downarrow + 5H_2O$$

Procedure

The ($-$) enantiomer is obtained by the procedure for the ($+$) enantiomer, substituting, however, the ($-$) form of the resolving agent[8] ($[\alpha]_D^{25} = -720°$) for the ($+$) form. The yield of ($-$),($-$) diastereomer is 4.2 g (84%). A 0.02% aqueous solution in a 1-dm tube gives $\alpha_D^{25} = -0.27°$, from which $[\alpha]_D^{25} = -1300°$.

Anal. Calcd. for $[Co(C_2O_4)(C_2H_8N_2)_2]_3[Cr(C_2O_4)_3]\cdot 6H_2O$: C, 23.52; H, 4.94; N, 13.72. Found: C, 23.50; H, 5.05; N, 13.84.

The yield of ($-$) enantiomer is 1.43 g or 78% based on $K_3[Cr(C_2O_4)_3]\cdot 3H_2O$ or 92% based on the diastereomer. A 0.02% aqueous solution in a 1-dm tube gives $\alpha_D^5 = -0.38°$, from which $[\alpha]_D^5 = -1900°$.

Anal. Calcd. for $K_3[Cr(C_2O_4)_3]\cdot H_2O$: C, 15.96; H, 0.45. Found: C, 15.95; H, 0.58.

The high optical density of even dilute solutions makes observation of the field difficult. The half-shade angle control on the Rudolph high-precision polarimeter should be opened to the maximum extent.

Properties

Although Werner[1] considered both enantiomers of potassium tris-(oxalato)chromate(III) to be monohydrates, Charonnat[11] and Delépine,[12] on the basis of isomorphism with the corresponding iridium salt, proposed that they are dihydrates. The (+) enantiomer has been described as bluish red in the dry, pure condition.[1,13] The optical enantiomers are less soluble in water than the racemic mixture. Because of the high rate of racemization in water, recrystallization has been reported to be accompanied by a considerable loss of optical rotation.[6,14–16] For the same reason, the optical rotation values reported by different workers are not directly comparable.[17]

Racemization also occurs in the solid state, but more slowly than in solution. It is faster and more extensive in the case of the hydrate at room temperature than in that of the anhydrous salt at 115°. The anhydrous salt can be stored up to 3 weeks *in vacuo* over phosphorus(V) oxide without loss of optical activity, but racemization occurs within 12 hr on heating the anhydrous salt in the presence of water vapor.

In aqueous solution potassium tris(oxalato)chromate(III) racemizes faster than the analogous cobalt compound but more slowly than the corresponding iron compound.[3] Many studies of the racemization rate have been made under various conditions.[1,6,18,19] The rate is dependent on the nature of the solvent; it is lowered by addition of organic solvents to aqueous solutions.[1,18] Both enantiomers have been the subject of numerous optical rotatory dispersion studies.[5,20–22] No exchange has been shown to occur between oxalate ion containing radioactive carbon and the tris(oxalato)-chromate(III) ion.[23]

References

1. A. Werner, *Berichte der Deutschen Chemischen Gesellschaft*, **45**, 3061 (1912).
2. G. B. Kauffman, L. T. Takahashi, and N. Sugisaka, *Inorg. Synth.*, **8**, 207 (1966).
3. W. Thomas and R. Fraser, *J. Chem. Soc.*, **123**, 2973 (1923).
4. F. P. Dwyer and E. C. Gyarfas, *Proceedings of the Royal Society of New South Wales*, **83**, 263 (1949).
5. F. M. Jaeger, *Rec. Trav. Chim.*, **38**, 142, 243 (1919).
6. C. H. Johnson and A. Mead, *Trans. Faraday Soc.*, **31**, 1621 (1935).
7. F. P. Dwyer and A. M. Sargeson, *J. Phys. Chem.*, **60**, 1331 (1956).
8. F. P. Dwyer, I. K. Reid, and F. L. Garvan, *J. Am. Chem. Soc.*, **83**, 1285 (1961).
9. J. C. Bailar, Jr., and E. M. Jones, *Inorg. Synth.*, **1**, 35 (1939).
10. W. G. Palmer, *Experimental Inorganic Chemistry*, Cambridge University Press, Cambridge, England, 1954, p. 386.
11. R. Charonnat, *Ann. Chim.*, [10] **16**, 5 (1931).
12. M. Delépine, *Bull. Soc. Chim. France*, [5] **1**, 1256 (1934).

13. L. Calderoni, *Boll. Chim. Farm.*, **71**, 517, 525 (1932).
14. C. H. Johnson, *Trans. Faraday Soc.*, **31**, 1612 (1935).
15. N. W. D. Beese and C. H. Johnson, *Trans. Faraday Soc.*, **31**, 1632 (1935).
16. E. Bushra and C. H. Johnson, *J. Chem. Soc.*, **1939**, 1937.
17. T. M. Lowry, *Bur. Stand. Miscellan. Publ.*, **118**, 1, 86 (1932).
18. G. K. Schweitzer and J. L. Rose, *J. Phys. Chem.*, **56**, 428 (1952).
19. W. D. Treadwell, G. Szabados, and E. Haimann, *Helv. Chim. Acta*, **15**, 1049 (1932).
20. J. P. Mathieu, *J. Chim. Phys.*, **33**, 78 (1936).
21. J. Lifschitz and E. Rosenbohm, *Zeitschrift für Wissenschaftliche Photographie, Photophysik und Photochemie* **19**, 198 (1920).
22. W. Kuhn and A. Szabo, *Z. Physik. Chem.*, **B15**, 59 (1932); W. Kuhn and K. Bein, *Z. Anorg. Allgem. Chem.*, **216**, 321 (1934).
23. F. A. Long, *J. Am. Chem. Soc.*, **61**, 570 (1939); **63**, 1353 (1941).

34.   DIFLUORODIOXOURANIUM(VI)

$$UO_2(NO_3)_2 \cdot 6H_2O + H_2O_2 \longrightarrow UO_4 \cdot 2H_2O + 2HNO_3 + 4H_2O$$

$$UO_4 \cdot 2H_2O + 3HF \longrightarrow H[UO_2F_3] \cdot 2H_2O + H_2O_2$$

$$H[UO_2F_3] \cdot 2H_2O \xrightarrow[150°]{\text{heated at}} UO_2F_2 + HF + 2H_2O$$

Submitted by M. C. CHAKRAVORTI* and MANJU CHOWDHURY*
Checked by P. G. ELLER† and R. J. KISSANE†

The fluoro compounds of uranium are of special interest in atomic energy programs. The compound, UO_2F_2, commonly called uranyl fluoride, is the best known oxyfluoride of hexavalent uranium and serves as a starting material in the synthesis of oxyfluoro or mixed ligand oxyfluoro uranium compounds. The anhydrous compound is usually prepared[1] by the action of gaseous hydrogen fluoride on UO_3 at temperatures of 300 to 500°. Other methods[1,2] involve reaction between anhydrous UO_2Cl_2 and anhydrous hydrofluoric acid at room temperature or by a high-temperature UO_3—F_2 reaction. In the method described here UO_2F_2 can be conveniently prepared in high yield using uranyl nitrate 6-hydrate, $UO_2(NO_3)_2 \cdot 6H_2O$ and aqueous hydrofluoric acid (20%). The advantages offered by this method are that it starts with the most commonly available compound of uranium, namely, $UO_2(NO_3)_2 \cdot 6H_2O$ and makes the use of anhydrous hydrogen fluoride or fluorine unnecessary. The method can be used by any worker in

*Department of Chemistry, Indian Institute of Technology, Kharagpur-721302, India.
†Los Alamos National Laboratory, Los Alamos, NM 87545.

any ordinary laboratory without the need of working with fluorine or anhydrous hydrogen fluoride at elevated temperature. The preparation of UO_2Cl_2 appeared earlier in *Inorganic Syntheses*.[3]

Procedure

■ **Caution.** *Hydrofluoric acid is toxic and highly corrosive. All operations with it should be carried out in an efficient fume hood. It should not be inhaled. Polyethylene gloves should be used while handling it.*

Ten grams (19.8 mmol) of $UO_2(NO_3)_2 \cdot 6H_2O$ is dissolved in 50 mL of water, and 0.3 mL of concentrated HNO_3 is added. The solution is heated to about 70° and 5 mL of hydrogen peroxide (30%) is added to the solution while stirring. A yellow precipitate of $UO_4 \cdot 2H_2O$ appears. After digesting for 0.5 hr on a steam bath the product is washed with water several times by decantation and is then filtered under suction. The precipitate is thoroughly washed with water and then transferred to a basin or dish made of polyethylene. It is dissolved in 12 mL (20%) hydrofluoric acid by stirring with a polyethylene stirrer. The solution is dried completely on a steam bath, and the product is finally dried in a desiccator over concentrated sulfuric acid for 7 days. A small dish containing pellets of caustic soda is also placed in the desiccator.* The dried product has the composition[4] $H[UO_2F_3] \cdot 2H_2O$.

Anal. Calcd. for $H[UO_2F_3] \cdot 2H_2O$: U, 65.4; F, 15.7. Found: U, 65.3; F, 15.8.

It is transferred into a platinum dish and heated at 150° for about 2 hr. The yield of UO_2F_2 is 5.5 g, 90%.

For the analysis of uranium, the sample is ignited in a platinum crucible and weighed as U_3O_8. Fluoride is determined by titration with a standard solution of thorium nitrate using sodium alizarinsulfonate as an indicator, after steam distillation of fluosilicic acid.[5]

Anal. Calcd. for UO_2F_2: U, 77.2; F, 12.3. Found: U, 76.9; F, 12.0.

Properties

The substance is bright yellow in color and is very soluble in water. It is soluble in ethanol, but insoluble in acetone and diethyl ether. It is mod-

---

*If the product dried on a steam bath is first subjected to pumping for 4 hr and then kept in a desiccator, as described, the time is reduced to 3 to 4 days. This time may be saved if $H[UO_2F_3] \cdot 2H_2O$ containing a little adsorbed HF is directly heated at 150°.

erately hygroscopic, and on standing in air, it absorbs water to form a dihydrate. (The checkers report that the product contained 0.59% H_2O by Karl Fischer titration.) In the IR spectra, a UO_2 band occurs at 1010 cm^{-1}. The crystal structure has been studied.[6,7]

References

1. J. J. Katz and E. Rabinowitch, *The Chemistry of Uranium*, Part 1, McGraw-Hill, New York, 1951, pp. 565–566.
2. J. C. Bailar, Jr. (Ed.), *Comprehensive Inorganic Chemistry*, Vol. 5, Pergamon, New York, 1973, p. 176.
3. J. A. Leary and J. F. Suttle, *Inorg. Synth.*, **5**, 148 (1957).
4. M. C. Chakravorti and N. Bandyopadhyay, *J. Inorg. Nucl. Chem.*, **34**, 2867 (1972).
5. G. Charlot and D. Bezier, *Quantitative Inorganic Analysis*, English translation by R. C. Murray, Methuen, London, 1957, p. 424.
6. W. H. Zachariasen, *Acta Cryst.*, **1**, 277 (1948).
7. N. V. Belov, *Doklady Akad. Nauk SSSR*, **65**, 677 (1949).

35. THE AMMONIUM CHLORIDE ROUTE TO ANHYDROUS RARE EARTH CHLORIDES—THE EXAMPLE OF YCl₃

$$12(NH_4)Cl + Y_2O_3 \longrightarrow 2(NH_4)_3[YCl_6] + 6NH_3 + 3H_2O \quad (1a)$$

$$6 (NH_4)Cl + 2YCl_3 \cdot 6H_2O \longrightarrow 2(NH_4)_3[YCl_6] + 12H_2O \quad (1b)$$

$$12(NH_4)Cl + 2Y \longrightarrow 2(NH_4)_3[YCl_6] + 6NH_3 + 3H_2 \quad (1c)$$

$$2(NH_4)_3[YCl_6] \longrightarrow \{(NH_4)[Y_2Cl_7] + 5(NH_4)Cl\}$$

$$\{(NH_4)[Y_2Cl_7] + 5(NH_4)Cl\} \longrightarrow 2YCl_3 + 6(NH_4)Cl \quad (2)$$

Submitted by GERD MEYER*
Checked by EDUARDO GARCIA† and JOHN D. CORBETT†

Among the methods for the preparation of anhydrous rare earth metal trichlorides, MCl_3 (M = La–Lu,Y,Sc),[1] the so-called *ammonium chloride route*[2] is probably the most popular and frequently used because it is inexpensive and straightforward even for large scale quantities. It was originally believed that the conversion of rare earths, M_2O_3, to trichlorides,

*Work performed at Institut für Anorganische und Analytische Chemie, Justus-Liebig-Universität, 6300 Giessen, Federal Republic of Germany. Present address: Institut für Anorganische Chemie, Universität Hannover, Callinstrasse 9, 3000 Hannover, Federal Republic of Germany.

†Department of Chemistry, Iowa State University, Ames, IA 50011.

MCl_3, is accomplished simply and directly by heating a mixture of rare earth oxides and excess ammonium chloride to a temperature of 200° or higher.[2] Excess ammonium chloride was thought to be necessary even recently[3] to prevent *hydrolysis*, that is, mainly the formation of oxychlorides, MOCl. Therefore, in the second step of the procedure the remaining ammonium chloride is then removed completely by heating in a vacuum at 300° to 320°.[2]

It was not recognized prior to 1982 that the *excess* ammonium chloride is not only necessary to prevent oxychloride formation but also essential in the early stages to form complex halides such as $(NH_4)_3[MCl_6]$ for M = Tb–Lu,Y,Sc, and $(NH_4)_2[MCl_5]$ for M = La–Gd.[4] This first step of the synthesis proceeds with reasonable speed at about 230°. The second step, in which the product of the first step is heated in a vacuum at 300°, is actually the decomposition of these complex chlorides and removal of any excess ammonium chloride. In some cases, the decomposition of $(NH_4)_3[MCl_6]$ passes through the intermediate $(NH_4)[Y_2Cl_7]$, for example, for M = Y,[4] but this has no effect on the purity of the final product, MCl_3. If insufficient $(NH_4)Cl$ is used in the first step, that is, not at least 12 mol $(NH_4)Cl$/mol M_2O_3, to convert all of the reactant to $(NH_4)_3[MCl_6]$ {or 10 mol where $(NH_4)_2[MCl_5]$ forms}, or if other reaction conditions like temperature or time are not observed carefully, the $(NH_4)_3[MCl_6]$ formed will react at about 300 to 330° with M_2O_3 to form MOCl,[5]

$$2(NH_4)_3[MCl_6] + 5M_2O_3 \longrightarrow 12MOCl + 6NH_3 + 3H_2O$$

The reaction of $(NH_4)_3[MCl_6]$ with water vapor must also be taken into account:[5,6]

$$(NH_4)_3[MCl_6] + H_2O \longrightarrow MOCl + 3(NH_4)Cl + 2HCl$$

Oxygen, for example, from dry air, seems not to react with $(NH_4)_3[MCl_6]$ or MCl_3 at the temperatures used here.[3] Except for this classical procedure for the syntheses of $(NH_4)_3[MCl_6]$ or $(NH_4)_2[MCl_5]$ [route eq. (1a)], a *wet* variant roughly described by eq. (1b) is even more straightforward for small scale quantities. The oxidation of the metal [variant eq. (1c)] is similar to the dry route eq. (1a) but seems less desirable since the metals are generally much more expensive than the sesquioxides. For a similar route to trichlorides from the metal and HCl gas, see ref. 7.

Procedure

Starting materials are commercially available rare earth oxides [Johnson-Matthey, Union Molycorp, Rhone-Poulenc] (generally M_2O_3 except for M = Pr, Tb where the compositions are approximately Pr_6O_{11} and Tb_4O_7)

or the hydrated chlorides $MCl_3 \cdot xH_2O$ ($x = 7$ for M = La–Nd and $x = 6$ for Sm–Lu,Y,Sc), or the metals themselves, preferably as powders, and ammonium chloride (Merck, Aldrich, etc.).

Step 1. The first step is the synthesis of the complex ammonium chlorides $(NH_4)_2[MCl_5]$ for M = La–Gd and $(NH_4)_3[MCl_6]$ for M = Tb–Lu,Y,Sc. This step is described here in three variations for the example of M = Y, but the syntheses of others can be designed very easily following the outlined procedure.

1a. For large scale quantities, for example, 50 g of Y_2O_3 and ≃ 150 g of $(NH_4)Cl$, the procedure as previously described is advisable.[2] Care should be taken that at least 12 mol of $(NH_4)Cl$, or better, 15 mol,[3] are used per mole of sesquioxide, Y_2O_3.

For small scale quantities, for example, 1 mmol Y_2O_3 (≡225.8 mg), a Pyrex ampule of about 12 mm o.d. is loaded with the reaction mixture of 225.8 mg of Y_2O_3 and 706.1 mg of $(NH_4)Cl$ [i.e., 12 mmol of $(NH_4)Cl$ + 10% excess], previously ground (in an agate mortar) preferably under dry conditions, that is, in a dry box. The neck of the ampule is pulled so that a capillary opening is produced as shown in Fig. 1.

The loaded ampule is then placed in a tubular furnace, and the temperature is slowly raised to 230–250° and held there for 10–20 hr. The progress of the reaction can be followed by droplets of water condensing in the capillary or by the alkaline reaction of the evolving ammonia with water. Excess $(NH_4)Cl$ is usually found to crystallize in the cooler parts of the ampule. The product $(NH_4)_3[YCl_6]$ is obtained in essentially quantitative yield.

1b. The wet route to $(NH_4)_3[YCl_6]$ starts either with $YCl_3 \cdot 6H_2O$ (commercially available as yttrium chloride), for example, 606.7 mg ≡ 2 mmol, or simply Y_2O_3 (225.8 mg ≡ 1 mmol), which is dissolved

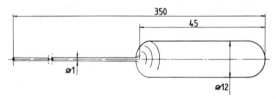

Fig. 1. Pyrex ampule with capillary opening as a container for reactions of $(NH_4)Cl$ with Y_2O_3 or Y metal. Proposed sizes are in millimeters. A shorter capillary (≈150 mm) but with a smaller opening (≈0.5 mm) is equally useful.

together with 353.0 mg $(NH_4)Cl$ ($\equiv 6$ mmol $+ 10\%$) in about 50 mL 12 M hydrochloric acid. When Y_2O_3 is used, heating to the boiling point is necessary to obtain a clear solution. This solution is then evaporated to dryness on a sand bath or a heating plate, or with an IR lamp. Slow evaporation is advisable to prevent spattering.

The dry residue, which is at least partly $(NH_4)_2[YCl_5 \cdot H_2O]$, is transferred to a corundum or glass boat of appropriate size (e.g., 10×60 mm) and inserted into the center of a Pyrex tube in a tubular furnace. A convenient apparatus has been described for the very similar synthesis of, for example, $K[Dy_2Cl_7]$.[8] Dry HCl gas is then passed for 1–2 days over the sample, which is heated to no higher than 230–250°. The product, $(NH_4)_3[YCl_6]$, should be completely soluble in water without cloudiness, which serves as a test for contamination by oxychloride, MOCl.

1c. The procedure of variant (1a) is followed with the Pyrex ampule loaded with, for example, 177.8 mg of yttrium metal powder ($\equiv 2$ mmol) and 706.1 mg $(NH_4)Cl$ ($\equiv 12$ mmol $+ 10\%$). The temperature must be raised to 270–300°.[9]

Step 2. The second step is the thermal decomposition of the previously synthesized complex ammonium chlorides, $(NH_4)_3[MCl_6]$ or $(NH_4)_2[MCl_5]$.

The $(NH_4)_3[YCl_6]$ obtained via one of the three routes just outlined is transferred in a dry box to a platinum, porcelain, or even glass crucible. This is then inserted into a wide glass tube (≈ 40 mm o.d.), which is fused to a water-cooled condenser (Fig. 2), or more simply to a long glass tube (≈ 200 mm), which may be joined to a trap with dry ice–acetone (usually not necessary) to prevent the attached vacuum line from being contaminated with $(NH_4)Cl$.

The temperature is raised slowly to above 300°, preferably 350–400°, under *dynamic* vacuum (oil rotary or, even better, mercury diffusion pump). A few hours is usually sufficient to ensure complete decomposition to YCl_3.

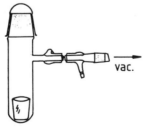

Fig. 2. Apparatus for the decomposition of, for example, $(NH_4)_3[YCl_6]$.

Properties

The rare earth trichlorides, MCl_3, are obtained as finely divided, hygroscopic powders that give broad lines on Debye or Guinier X-ray patterns. This means that the products are of only poor crystallinity and therefore highly reactive, especially those that have the YCl_3-type structure ($DyCl_3$–$LuCl_3$, YCl_3). Others crystallize with the UCl_3-type ($LaCl_3$–$GdCl_3$) and $TbCl_3$ with the $PuBr_3$-type structure. Indicative for YCl_3 are strong X-ray lines at d = 2.740 (100), 2.745 (47), 5.682 (57), 6.012 Å (84) with relative intensities on a 1 to 100 scale given in parentheses. For further crystallographic details see, for example, ref. 10. Principal impurities are oxychlorides, MOCl. These are obtained in the PbFCl (for M = La–Dy) and the YOF–SmSI structure types under the conditions just outlined.[6] Especially the reaction of $(NH_4)_3[YCl_6]$ with Y_2O_3 to YOCl is troublesome and pertains to the use of an insufficient quantity of ammonium chloride in the first step. Indicative for the presence of YOCl is especially the low-angle reflection at d = 9.32 Å (100). For a very high purity product (YCl_3) one or more subsequent sublimations in a completely tantalum apparatus should be carried out.[7] When sufficient $(NH_4)Cl$ was used, however, this has been found not to be necessary for most applications. The trichlorides, MCl_3, should be stored in sealed Pyrex ampules under dry inert gas (Ar, N_2) and handled only under dry conditions (dry box).

References

1. M. D. Taylor, *Chem. Rev.*, **62**, 503 (1962).
2. J. B. Reed, B. S. Hopkins, and L. F. Audrieth, *Inorg. Synth.*, **1**, 28 (1939).
3. Y. S. Kim, F. Planinsek, B. J. Beaudry, and K. A. Gschneidner, Jr., in *The Rare Earths in Modern Science and Technology*, Vol. 2, G. M. McCarthy, J. J. Rhyne, and H. B. Silber (eds.), Plenum, New York, 1980, p. 53.
4. G. Meyer and P. Ax, *Mater. Res. Bull.*, **17**, 1447 (1982).
5. G. Meyer and Th. Staffel, *Z. Anorg. Allg. Chem.*, **532**, 31 (1986).
6. E. Garcia, J. D. Corbett, J. E. Ford, and W. J. Vary, *Inorg. Chem.*, **24**, 494 (1985).
7. J. D. Corbett, *Inorg. Synth.*, **22**, 39 (1983).
8. G. Meyer, *Inorg. Synth.*, **22**, 1 (1983).
9. G. Meyer, Th. Staffel, S. Dötsch, and Th. Schleid, *Inorg. Chem.*, **24**, 3504 (1985).
10. E. Fluck (ed.) *Gmelins Handbuch der Anorganischen Chemie*, 8th ed., Part C 5, Springer-Verlag, Berlin, 1977.

Chapter Five

TRANSITION METAL ORGANOMETALLIC COMPOUNDS

36. BIS(PHOSPHINE) DERIVATIVES OF IRON PENTACARBONYL AND TETRACARBONYL (TRI-*tert*-BUTYLPHOSPHINE)IRON(0)

Submitted by MICHAEL J. THERIEN† and WILLIAM C. TROGLER*
Checked by ROSALICE SILVA† and MARCETTA Y. DARENSBOURG†

Bis(phosphine) derivatives of pentacarbonyliron are starting materials for the synthesis of several organometallic iron complexes.[1-7] Iron carbonyl phosphine complexes have attracted attention[8-11] because of their relevance to photochemical catalysis of olefin hydrosilation. Though $Fe(CO)_3(PR_3)_2$ complexes are used widely in organotransition metal chemistry, an efficient preparation of these compounds has not been reported. Clifford and Mukherjee[12] describe two methods for the synthesis of tricarbonyl-bis(triphenyphosphine)iron(0). They report that direct reaction between $Fe_3(CO)_{12}$ and triphenylphosphine in THF solvent gives a mixture of $Fe(CO)_3[P(C_6H_5)_3]_2$ (27%) and $Fe(CO)_4[P(C_6H_5)_3]$ (34%). The second

*Department of Chemistry, D-006, University of California at San Diego, La Jolla, CA 92093.
†Department of Chemistry, Texas A&M University, College Station, TX 77843.

method [the reaction between $Fe(CO)_5$ and $P(C_6H_5)_3$ in cyclohexanol] gives both $Fe(CO)_3[P(C_6H_5)_3]_2$ and $Fe(CO)_4[P(C_6H_5)_3]$ in 15% yield. Again, the mono- and bis-substituted compounds have to be separated by vacuum sublimation.

A better synthesis (89% yield) of $Fe(CO)_3(PPh_3)_2$ is reported[13] from $[PPN]_2[Fe_4(CO)_{13}]$, where PPN^+ = bis(triphenylphosphine)iminium. The CoX_2 (X = Cl, Br, I) catalyzed substitution of CO in $Fe(CO)_5$ is reported[14] to yield $Fe(CO)_4L$ species in 15 to 99% yield and $Fe(CO)_3(PPh_3)_2$ was prepared (net 62% yield) from $Fe(CO)_5$ in a two-step procedure that requires a chromatographic separation. Strohmeier and Muller[15] report that irradiation of $Fe(CO)_5$ in the presence of several phosphines produces $Fe(CO)_3L_2$ and $Fe(CO)_4L$ complexes in yields that range from 13% for the synthesis of $Fe(CO)_3[P(n\text{-}Bu)_3]_2$ to 35% for $Fe(CO)_3[P(c\text{-}C_6H_{11})_3]_2$. For some of the compounds synthesized, vacuum sublimation is necessary to separate the $Fe(CO)_3L_2$ species from $Fe(CO)_4L$. The one-step photochemical procedure we report here employs cyclohexane as a solvent. That enables unreacted phosphine, $Fe(CO)_5$, and $Fe(CO)_4L$ to remain in solution while pure $Fe(CO)_3L_2$ precipitates. It is essential that the phosphines used in these reactions be free of phosphine oxides, which labilize[16] CO and yield products other than $Fe(CO)_3(PR_3)_2$ complexes.

General Procedure

Method 1. For large scale reactions, 300 to 450 mL of cyclohexane is heated at reflux over sodium, distilled, and transferred under N_2 into a photochemical reaction vessel (500 mL) that is fitted around a 450-W Hanovia mercury arc lamp contained in a water cooled quartz immersion well (see Fig. 1). This apparatus is available from Ace Glass.* A septum-capped sidearm permits the system to be flushed with N_2 through a syringe needle. The carbon monoxide evolved during the reaction is collected by venting the air-tight irradiation vessel through a mineral oil bubbler. The outlet of the bubbler is connected, via Tygon tubing, to release gas into the bottom of an inverted 2000- or 3000-mL graduated cylinder that is filled with water and contained in a partially filled 5000-mL beaker. The volume of CO gas produced is measured as it displaces water in the graduated cylinder. The oil bubbler prevents exposure of the reaction solution to water vapor and serves as a safety valve if a slight back pressure develops. Two to 10 mL of $Fe(CO)_5$ (152 to 760 mmol) is introduced by syringe into the cyclohexane

---

*Ace Glass Co., P.O. Box 688, 1430 Northwest Blvd., Vineland, NJ 08360.

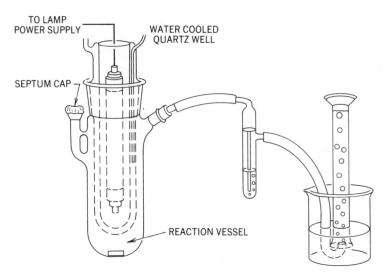

Fig. 1. Experimental setup for large scale photochemical syntheses.

solvent along with 3 to 5 equivalents of phosphine through the septum capped sidearm. Photolysis is begun with stirring, and the volume of gas evolved is monitored as light yellow $Fe(CO)_3L_2$ precipitates during the reaction. Irradiation is stopped when the volume of CO gas evolved equals the theoretical amount or when the reaction mixture ceases to evolve CO. At this point one must carefully remove the immersion well from the center of the reaction vessel (under an N_2 flush) without dislodging solid decomposition products that adhere to the surface of the inner quartz well. The solution is filtered under N_2, and the recovered $Fe(CO)_3(PR_3)$ is dried under vacuum. The volume of the filtrate is reduced by 50% and then filtered to yield more $Fe(CO)_3(PR_3)_2$. The recovered $Fe(CO)_3(PR_3)_2$ complexes give satisfactory elemental analyses and IR spectra and require no further purification.

Method 2. A convenient method for small scale preparations of these compounds uses Schlenk techniques. To a 100-mL quartz Schlenk tube is added 0.5 to 1.0 mL (38. to 76. mmol) of $Fe(CO)_5$, 3 to 5 equivalents of phosphine, and 80 mL of dry deoxygenated cyclohexane. The Schlenk tube is placed adjacent to the mercury arc lamp contained in the quartz immersion well. The sidearm of the Schlenk flask is attached to N_2 flushed Tygon tubing, which is connected to the bubbler and an inverted water-filled graduated cylinder (500 mL) contained in a partially filled beaker.

It is best to let the photochemical reaction proceed for a few minutes before opening the stopcock. This allows a pressure of CO to build up; this forces air in the Tygon tubing to the other side of the bubbler when the stopcock is opened. When complete, the reaction is worked up as in the large scale procedure just described. The checkers found reduced yields when the reactions were conducted on a smaller scale without the specified excess of phsophine ligand.

■ **Caution.**   *The compound Fe(CO)₅ is toxic and should be used only in a fume hood. Care should be used when handling trimethyl- and triethylphosphine since they are toxic and ignite readily. The UV lamp will cause severe eye damage or blindness if it is viewed without UV-protective goggles. It should be concealed from sight in a light-tight box during photolysis.*

A.   TRICARBONYLBIS(TRIPHENYLPHOSPHINE)IRON(0)

$$Fe(CO)_5 + 2P(C_6H_5)_3 \xrightarrow{h\nu} Fe(CO)_3[P(C_6H_5)_3]_2 + 2CO$$

Method 1 is used for the reaction between 10 mL of Fe(CO)₅ and 95 g of P(C₆H₅)₃. After 15 hr of irradiation, 2.95 L of gas is evolved, and the reaction is stopped. The first filtration yields 28.1 g of product. Reducing the volume by 50% yields an additional 5.2 g of Fe(CO)₃[P(C₆H₅)₃]₂. Yield: 33.3 g, 66%.

Anal.   Calcd. for Fe(CO)₃[P(C₆H₅)₃]₂: C, 70.50; H, 4.52; P, 9.34. Found: C, 70.54; H, 4.98; P, 9.23.

B.   TRICARBONYLBIS(TRICYCLOHEXYLPHOSPHINE)IRON(0)

$$Fe(CO)_5 + 2P(c\text{-}C_6H_{11})_3 \xrightarrow{h\nu} Fe(CO)_3[P(c\text{-}C_6H_{11})_3]_2 + 2CO$$

The photochemical reaction between 0.90 mL of Fe(CO)₅ (6.84×10^{-3} mol) and 9.3 g (3.32×10^{-2} mol) of P(c-C₆H₁₁)₃ in 300 mL of cyclohexane (Method 1) went to completion in 6 hr with the evolution of 320 mL of gas. After the initial filtration, 2.65 g of product is recovered. Reducing the volume of cyclohexane and refrigerating the solution yields an additional 0.75 g of product. Yield: 3.4 g; 71%.

Anal.   Calcd. for Fe(CO)₃[P(c-C₆H₁₁)₃]₂: C, 66.86; H, 9.43; P, 8.86. Found: C, 67.45; H, 9.52; P, 8.50.

C. TRICARBONYLBIS(TRI-*n*-BUTYLPHOSPHINE)IRON(0)

$$Fe(CO)_5 + 2P(n\text{-}Bu)_3 \xrightarrow{h\nu} Fe(CO)_3[P(n\text{-}Bu)_3]_2 + 2CO$$

Using Method 1, a mixture of 5 mL of $Fe(CO)_5$ (3.83×10^{-2} mol) and 21.2 g of $P(n\text{-}Bu)_3$ (1.04×10^{-1} mol) is irradiated. The reaction is complete after 12 hr. The initial filtration yields 3.7 g of product. After reducing the volume of the solution, an additional 2.2 g is obtained. Yield: 5.9 g; 28%.

Anal. Calcd. for $Fe(CO)_3[P(n\text{-}Bu)_3]_2$: C, 59.58; H, 9.93; P, 11.40. Found: C, 59.32; H, 9.86; P, 11.15.

D. TRICARBONYLBIS(TRIMETHYLPHOSPHINE)IRON(0)

$$Fe(CO)_5 + 2PMe_3 \xrightarrow{h\nu} Fe(CO)_3(PMe_3)_2 + 2CO$$

To 1 mL of $Fe(CO)_5$ (7.61×10^{-3} mol) was added 2.76 g of PMe_3 (3.63×10^{-2} mol) in 80 mL of dry cyclohexane (Method 2). The reaction vessel is cooled in a quartz beaker containing an ice–salt mixture to avoid loss of volatile PMe_3 during the irradiation even though the cooling bath scatters much of the light. After 5 hr the reaction is complete. The initial filtration yields 1.23 g of product. After reducing the volume of solvent by 50%, an additional 0.54 g is recovered. Yield: 1.77 g, 80%.

Anal. Calcd. for $Fe(CO)_3[P(CH_3)_3]_2$: C, 37.02; H, 6.21; P, 21.21. Found: C, 36.65; H, 6.27; P, 21.13.

E. TETRACARBONYLTRI-*tert*-BUTYLPHOSPHINEIRON(0)

$$Fe(CO)_5 + P(t\text{-}Bu)_3 \xrightarrow{h\nu} Fe(CO)_4[P(t\text{-}Bu)_3]$$

The reaction between 1.3 mL of $Fe(CO)_5$ (9.69×10^{-3} mol) and 6.31 g of $P(t\text{-}Bu)_3$ in 350 mL of dry cyclohexane (Method 1) produces 475 mL of CO after 8 hr of irradiation. The reaction mixture is filtered to yield 1.51 g of yellow product. An additional 1.05 g of solid is obtained by reducing the volume of solution. The combined yield (2.56 g) proves to be exclusively the monosubstituted product. Apparently, the conditions of the reaction do not allow two bulky tri-*tert*-butylphosphine ligands to replace two carbonyl groups on $Fe(CO)_5$. Yield: 2.56 g; 72%.

Anal. Calcd. for $Fe(CO)_4[P(t\text{-}Bu)_3]$: C, 51.91; H, 7.35; P, 8.37. Found: C, 51.43; H, 7.23; P, 8.32.

Properties

Both the $Fe(CO)_3(PR_3)_2$ complexes and $Fe(CO)_4[P(t\text{-}Bu)_3]$ are soluble in organic solvents such as CH_2Cl_2, benzene, toluene, THF, acetone, and hot heptane or cyclohexane, with the degree of solubility varying with the type of phosphine (the PPh_3 derivative is least soluble). All the complexes are air stable in the solid state. Prolonged exposure to light darkens the surfaces of these compounds. When left in solution for long periods of time and exposed to air, the compounds decompose slowly; those complexes containing small phosphine ligands seem to be the most sensitive. These compounds should be kept cold for long term storage. Physical properties are listed in the following table.

Compound	mp (°)	ν_{CO} (cm^{-1})^a	^{31}P (ppm)^b
$Fe(CO)_4[P(t\text{-}Bu)_3]$	170 dec	2040, 1960, 1920	126.9
$Fe(CO)_3(PMe_3)_2$	195	1863	42.4
$Fe(CO)_3[P(n\text{-}Bu)_3]_2$	55	1855	66.0
$Fe(CO)_3[P(c\text{-}C_6H_{11})_3]_2$	228	1846	89.1
$Fe(CO)_3[PPh_3]_2$	272	1878	78.2

^{a}Solution IR spectra used CH_2Cl_2 as the solvent.
^{b}NMR spectra were recorded on a Nicolet 200-MHz instrument in $CDCl_3$ solvent. Chemical shifts are parts per million (ppm) downfield from 85% H_3PO_4.

References

1. P. K. Baker, N. S. Connelly, B. M. R. Jones, J. P. Maher, and K. R. Somers, *J. Chem. Soc., Dalton Trans.*, **1980**, 579.
2. W. E. Carroll and F. J. Lalor, *J. Chem. Soc., Dalton Trans.*, **1973**, 1754.
3. G. R. Crooks and B. F. G. Johnson, *J. Chem. Soc. (A)*, **1968**, 1238.
4. R. K. Kummer and W. A. G. Graham, *Inorg. Chem.*, 1208, (1968).
5. A. Davison, W. McFarlane, L. Pratt, and G. Wilkinson, *J. Chem. Soc.*, **1962**, 3653.
6. K. Farmey and M. Kilner, *J. Chem. Soc. (A)*, **1970**, 634.
7. M. J. Therien, C-L. Ni, F. C. Anson, J. G. Osteryoung, and W. C. Trogler, *J. Am. Chem. Soc.*, **108**, 4037, (1986).
8. R. D. Sanner, R. G. Austin, M. S. Wrighton, W. D. Honnick, and C. U. Pittman, *Inorg. Chem.*, **18**, 928, (1979).
9. D. K. Liu, C. G. Brinkley, and M. S. Wrighton, *Organometallics*, **3**, 1449, (1984).
10. J. L. Graff, R. D. Sanner, and M. S. Wrighton, *Organometallics*, **1**, 837, (1982).
11. D. K. Liu, M. S. Wrighton, D. R. McKay, and G. R. Maciel, *Inorg. Chem.*, **23**, 212, (1984).
12. A. F. Clifford and A. K. Mukherjee, *Inorganic Synth.*, **8**, 184, (1966).
13. S. B. Butts and D. F. Shriver, *J. Organomet. Chem.*, **169**, 191, (1979).
14. M.O. Albers and N. J. Coville, *J. Organomet. Chem.*, **17**, 385, (1981).
15. W. Strohmeier and F. J. Muller, *Chem. Ber.*, **102**, 3613, (1969).
16. D. J. Darensbourg, M. Y. Darensbourg, and N. Walker, *Inorg. Chem.*, **20**, 1918, (1981).

37. TRICARBONYL BIS(*N,N*-DIALKYLCARBAMODITHIOATE)TUNGSTEN(II)

$$W(CO)_6 + Br_2 \longrightarrow WBr_2(CO)_4 + 2CO$$
$$WBr_2(CO)_4 + 2NaS_2CNR_2 \longrightarrow W(CO)_3[S_2CNR_2]_2 + CO + 2NaBr$$

Submitted by J.A. BROOMHEAD,* J. BUDGE,† W. PIENKOWSKI,* and
C.G. YOUNG*
Checked by T. L. TONKER‡ and J. L. TEMPLETON‡

There is renewed interest in the fundamental chemistry of tungsten arising from its congener relationship to molybdenum; an important element in both catalysis and metalloenzyme processes. The title complex has been prepared by Templeton and Ward[1] from $W(CO)_4I_2$ and the crystal and molecular structure reported for the dimethylcarbamodithioate analog. Their method requires a chromatographic purification step, which is not necessary in the following procedure using bromine as the oxidant. This method is similar to that described previously[2] for $Mo(CO)_2(S_2CNEt_2)_2$ and $Mo(CO)_3(S_2CNEt_2)_2$ and may be used with slight modifications for the corresponding dimethyl-, diisopropyl-, benzyl-, pyrrolidyl- and dicyclohexylcarbamodithioate ligands.

■ **Caution.** *Carbon monoxide is liberated in this reaction and it should be performed in a well-ventilated fume hood.*

Procedure

(All operations are carried out under nitrogen on a Schlenk line.) Tungsten hexacarbonyl (Fluka AG) (4 g, 0.011 mol) is suspended in deoxygenated dichloromethane (60 mL) in a Schlenk flask connected to a gas bubbler outlet and fitted with a rubber septum. The mixture is cooled at $-78°$ in acetone–dry ice and stirred magnetically while bromine (0.58 mL, 0.011 mol) is added by injection. After slight warming, there is vigorous gas evolution **(CARE!)** and a red-brown solution is formed. The solvent is removed on the vacuum line at room temperature and the vessel is alternatively filled with nitrogen and evacuated several times to ensure complete removal of bromine. The orange-brown residue is dissolved in methanol (30 mL) and further gas evolution takes place to give a brown solution containing dibromotetracarbonyltungsten(II). This is filtered directly into

*Faculty of Science, Australian National University, Canberra, A.C.T. 2601. Australia.
†BP America, Independence, OH 44131-5595.
‡Department of Chemistry, University of North Carolina at Chapel Hill, Chapel Hill, NC 27514.

a stirred slurry of $NaS_2CNEt_2 \cdot 3H_2O$ (5.17 g, 0.023 mol) in methanol (30 mL), (or equimolar amounts of the corresponding NaS_2CNR_2 salts). The crude orange-brown solid is filtered off, washed with methanol (2 × 15 mL) and pumped dry. The product is dissolved in dichloromethane (~15 mL) and precipitated as orange crystals by the addition of methanol (~60 mL). Yield: 4.5 g (73%).

Anal. Calcd. for $C_{13}H_{20}N_2O_3S_4W$: C, 27.7; H, 3.6; N, 5.0; S, 22.7. Found: C, 27.7; H, 3.8; N, 4.9; S, 22.5.

Properties

The tricarbonylbis(N,N-dialkylcarbamodithioate)tungsten(II) complexes are orange crystalline solids that may be stored indefinitely under nitrogen. Loss of one carbonyl ligand occurs on heating at reflux in methanol and the corresponding dicarbonyl complexes are formed. The tricarbonyl complexes are soluble in nonpolar solvents. For $W(CO)_3(S_2CNEt_2)_2$ the characteristic strong IR bands (for CsI discs) occur at 2020, 1937, 1917, 1902, and 1886 cm^{-1} ($\nu_{C \equiv O}$) and 1508 cm^{-1} ($\nu_{C \equiv N}$). The ^{13}C NMR spectrum at $-50°$ shows only one carbonyl resonance and equivalent diethylcarbamodithioate resonances indicating fluxionality in these ligands. The ^1H NMR and IR spectra of the dimethyldithiocarbamate derivative and of the dicarbonyl complexes have been reported.[3]

References

1. J. L. Templeton and B. C. Ward. *Inorg Chem.*, **19**, 1753 (1980).
2. J. A. Broomhead, J. Budge, and W. Grumley. *Inorg. Synth.*, **16**, 235 (1976).
3. J. A. Broomhead and C. G. Young. *Aust. J. Chem.*, **35**, 277 (1982).

38. CYCLOPENTADIENYLBIS(TRIMETHYLPHOSPHINE) AND CYCLOPENTADIENYLBIS(TRIMETHYLPHOSPHITE) COMPLEXES OF Co and Rh

Submitted by H. WERNER,* R. FESER,* V. HARDER,* W. HOFMANN,* and H. NEUKOMM*
Checked by W. D. JONES†

The title complexes, $CpM(PMe_3)_2$ and $CpM[P(OMe)_3]_2$, are electron-rich half-sandwich complexes that have been demonstrated to be valuable pre-

*Institut für Anorganische Chemie der Universität, Am Hubland, D-8700 Würzburg, Federal Republic of Germany.
†Department of Chemistry, The University of Rochester, Rochester, NY 14627.

cursors to a large number of organometallic derivatives.[1] They behave as Lewis bases and react with a wide variety of electrophiles E or EX to form new metal–element bonds. They have also been shown to be valuable starting materials for the syntheses of heterometallic di- and trinuclear complexes via their reactions with unsaturated transition metal compounds.[1] The only viable syntheses of these compounds now known are those reported here.

A. (η⁵-CYCLOPENTADIENYL)BIS-(TRIMETHYLPHOSPHINE)RHODIUM(I)[2]

$$[(C_8H_{14})_2RhCl]_2 + 4PMe_3 + 2LiC_5H_5 \longrightarrow$$
$$2Rh(\eta\text{-}C_5H_5)(PMe_3)_2 + 2LiCl + 2C_8H_{14}$$

Procedure

In a 125-mL Schlenk tube, equipped with a nitrogen inlet and a magnetic stirring bar, 3.7 g (5.15 mmol) $[(C_8H_{14})_2RhCl]_2$ (ref. 3) is dissolved in 40 mL of THF, freshly distilled over Na and benzophenone. The solution is treated dropwise with 2.1 mL (21.0 mmol) PMe$_3$ and then stirred for 1 hr at room temperature. A 1.01 g (14.0 mmol) quantity of LiC$_5$H$_5$, freshly prepared from equimolar amounts of *n*-BuLi and C$_5$H$_6$ in hexane,[4] is added and the reaction mixture is stirred for 2 hr at room temperature. The suspension is filtered, and the red-brown filtrate is evaporated to dryness under reduced pressure. The solid residue is extracted with pentane (2 × 20 mL). The pentane solution is filtered, and the filtrate is concentrated *in vacuo* to ~5 mL. After cooling to $-78°$, red-brown air-sensitive crystals are obtained, which must be stored under thoroughly purified nitrogen or argon (preferably in a refrigerator). Yield: 2.37 g (72%), mp 85°.

Anal. Calcd. for $C_{11}H_{23}P_2Rh$: C, 41.26; H, 7.24; P, 19.35; Rh, 32.13; MW, 320.2. Found: C, 41.04; H, 7.09; P, 19.62; Rh, 32.38; MW, 320 (mass spectroscopy).

Note: In the original procedure,[2] the cycloocta-1.5-diene complex $[C_8H_{12}RhCl]_2$ was used as starting material. The present method avoids the isolation of the intermediate $[Rh(PMe_3)_4]Cl$.

Properties

(η-Cyclopentadienyl)bis(trimethylphosphine)rhodium(I) is a red-brown crystalline product that remains unchanged when stored under an inert atmosphere at 0–10°. It is soluble in hydrocarbon solvents but decomposes in chloroform and carbon tetrachloride. In methylene dichloride, an oxidative addition reaction occurs to give the cation $[Rh(\eta\text{-}C_5H_5)(PMe_3)_2\text{-}CH_2Cl]^+$, which can be isolated as the PF_6^- salt.[5] The ^1H NMR spectrum of the title compound (60 MHz, benzene-d_6, δ in ppm downfield from TMS) shows two signals at: 5.27 (triplet, $J_{PH} = 0.6$ Hz, C_5H_5 protons) and 1.16 (doublet of virtual triplets, $J_{RhH} = 1.2$ Hz, $N = 8.4$ Hz, PMe_3 protons).

The complex is a useful starting material for the preparation of other cyclopentadienylrhodium complexes, for example, $[Rh(\eta\text{-}C_5H_5)(PMe_3)_2R]\text{-}PF_6$ (R = H, Me, Et, COMe, COPh, GeMe$_3$, SnMe$_3$, Cl, Br, I),[2] $[Rh(\eta\text{-}C_5H_5)_2(PMe_3)_2(CH_2PMe_3)]I_2$,[5] $[Rh(\eta\text{-}C_5H_5)(PMe_3)(CH_2PMe_3)I]I$,[5] and $[Rh(\eta\text{-}C_5H_5)(PMe_3)_2(Al_2Me_4Cl_2)]$.[6]

B. (η^5-CYCLOPENTADIENYL)BIS-(TRIMETHYLPHOSPHINE)COBALT(I)[7]

$$2CoCl_2 + Mg + 6PMe_3 \longrightarrow 2CoCl(PMe_3)_3 + MgCl_2$$

$$CoCl(PMe_3)_3 + LiC_5H_5 \longrightarrow Co(\eta\text{-}C_5H_5)(PMe_3)_2 + PMe_3 + LiCl$$

The original procedures for $CoCl(PMe_3)_3$ (ref. 8) and $C_5H_5Co(PMe_3)_2$ (ref. 9) were modified as described next.

Procedure

(a) $CoCl(PMe_3)_3$. In a 125-mL Schlenk tube, equipped with a nitrogen inlet, a solution of 7 mL (73.7 mmol) of PMe$_3$ (ref. 10) in 50 mL of THF, freshly distilled over Na and benzophenone, is treated with 2.0 g (15.4 mmol) of anhydrous CoCl$_2$, 600 mg of Mg turnings, freshly cut, and 50 mg (0.39 mmol) of anthracene. The reaction mixture is placed for 20 min in an ultrasonic bath, which leads to a color change from violet to dark brown. The solvent is removed *in vacuo,* and the brown solid residue is extracted with ether (~3 × 20 mL). The ether solution (which is *very* air sensitive), together with 2.6 mL (27.3 mmol) of PMe$_3$,[10] is added to a Schlenk tube, equipped with a nitrogen inlet, and containing 1.4 g (10.8 mmol) of anhydrous CoCl$_2$. The reaction mixture is stirred for 2 hr at room temperature. The blue precipitate formed is filtered off, washed with ether, and dried *in vacuo.* The product must be stored under thoroughly purified nitrogen or argon at low temperature. Yield: 5.49 g (65%).

(b) $Co(\eta\text{-}C_5H_5)(PMe_3)_2$. In a 125-mL Schlenk tube, equipped with a nitrogen inlet and a magnetic stirring bar, 5.96 g (18.5 mmol) of $CoCl(PMe_3)_3$ is dissolved in 50 mL of THF, freshly distilled over Na and benzophenone, and the resulting solution cooled to $-50°$. To the solution, a quantity of 1.4 g (19.5 mmol) of LiC_5H_5 is added in small portions. The reaction mixture is stirred for 30 min at $-50°$ and then slowly warmed to room temperature. While keeping the temperature at 20–25°, the solvent is removed *in vacuo*, and the brown solid residue is extracted with hexane (3 × 20 mL). The hexane solution is filtered and concentrated *in vacuo* to ~5 mL. After cooling at $-78°$ dark brown air-sensitive crystals are isolated, which must be stored under thoroughly purified nitrogen or argon (preferably at low temperature). Yield: 4.76 g (93%), mp 55–57°.

Anal. Calcd. for $C_{11}H_{23}CoP_2$: C, 47.85; H, 8.34; Co, 21.36; P, 22.45; MW, 275.9. Found: C, 47.87; H, 8.12; Co, 21.10; P, 22.28; MW, 276 (mass spectrometry).

Properties

(η-Cyclopentadienyl)bis(trimethylphosphine)cobalt(I) is a dark brown crystalline product that remains unchanged when stored under an inert atmosphere at 0 to 10°. It is soluble in hydrocarbon solvents but decomposes rapidly in chloroform. In methanol, slow reaction takes place that leads to the formation of the cation $[Co(\eta\text{-}C_5H_5)(PMe_3)_2H]^+$. This cation (with PF_6^- as the anion) is formed more easily from $[Co(\eta\text{-}C_5H_5)(PMe_3)_2]$ and NH_4PF_6. The ^1H NMR spectrum of the title compound (60 MHz, benzene-d_6, δ in ppm downfield from TMS) shows two signals at: 4.51 (triplet, $J_{PH} = 1.4$ Hz, C_5H_5 protons) and 1.07 (virtual triplet, $N = 7.6$ Hz, PMe_3 protons).

The complex is a useful starting material for the preparation of other cyclopentadienylcobalt complexes, for example, $[Co(\eta\text{-}C_5H_5)(PMe_3)_2R]X$ (R = H, Me, Et, COMe, COPh, $SnCl_3$, $SnMe_3$, $SnPh_3$; X = I or PF_6),[9,11] $[Co(\eta\text{-}C_5H_4R)(PMe_3)_2]$ (R = $CHMe_2$, CMe_3, CMe_2Et),[12] $[Co(\eta\text{-}C_5H_5)(PMe_3)(\eta^2\text{-}CS_2)]$,[13] $[Co(\eta\text{-}C_5H_5)(PMe_3)(\eta^2\text{-}CSSe)]$,[14] $[Co(\eta\text{-}C_5H_5)(PMe_3)(\eta^2\text{-}CSe_2)]$,[14] $[Co(\eta\text{-}C_5H_5)(PMe_3)(CE)]$ (E = S, Se),[14] $[Co(\eta\text{-}C_5H_5)(PMe_3)(CNR)]$ (R = Me, CMe_3, Ph),[15] $[Co(\eta\text{-}C_5H_5)(PMe_3)(\eta^2\text{-}S_5)]$,[16] $[(\eta\text{-}C_5H_5)(PMe_3)Co(\eta\text{-}CO)_2Mn(CO)(\eta\text{-}C_5H_4Me)]$.[17] The carbondisulfide-cobalt complex has further been used for the synthesis of Co_3 clusters containing a bridging thiocarbonyl ligand.[18]

C.   (η⁵-CYCLOPENTADIENYL)BIS-
(TRIMETHYLPHOSPHITE)COBALT(I)[19]

$$Co(\eta\text{-}C_5H_5)_2 + 2P(OMe)_3 \longrightarrow [Co(\eta\text{-}C_5H_5)(P(OMe)_3)_2] + \{C_5H_5\}$$

Procedure

In a 20-mL Schlenk tube, equipped with a nitrogen inlet, a reflux condenser, and a magnetic stirring bar, 2.09 g (11.06 mmol) of $Co(\eta\text{-}C_5H_5)_2$ (ref. 20) and 7 mL (59.4 mmol) of $P(OMe)_3$ are heated under reflux for 3 days. After cooling to room temperature, excess of $P(OMe)_3$ is removed *in vacuo* (~15 torr), and the oily residue is distilled at ~149° and ~10^{-4} torr using a short path distillation apparatus. The distilled product is dissolved in pentane (~10 mL), and the pentane solution is concentrated *in vacuo*. Cooling to ~ −30° gives red-brown air-sensitive crystals that are filtered off, washed with small amounts of cold pentane, and dried *in vacuo*. Yield: 2.2 g (56%), mp 38°.

Anal. Calcd. for $C_{11}H_{23}CoO_6P_2$: C, 35.50; H, 6.23; Co, 15.84; MW, 372.2. Found: C, 35.39; H, 6.01; Co, 15.87; MW, 372 (mass spectrometry).

Properties

(η-Cyclopentadienyl)bis(trimethylphosphite)cobalt(I) is a red-brown crystalline low-melting product that is air sensitive and should be stored under an inert atmosphere at 10–20°. It is soluble in hydrocarbon solvents, but decomposes in chloroform and carbon tetrachloride. The ^1H NMR spectrum (60 MHz, acetone-d_6, δ in ppm downfield from TMS) shows two signals at: 4.63 (singlet, C_5H_5 protons) and 3.48 [virtual triplet, N = 12.0 Hz, $P(OMe)_3$ protons]. The UV spectrum (in hexane, λ_{max} in cm^{-1}) shows three maxima at: 23 920 (log ε 2.7), 38 910 (log ε 4.7), and 43 860 (log ε 5.0).

   The complex is a good nucleophile and reacts with acids (e.g., CF_3CO_2H) and methyl iodide to form the corresponding salts of the cations [Co(η-C_5H_5)(P(OMe)$_3$)$_2$R]$^+$ (R = H, Me).[21] Thermolysis gives the "supersandwich" complex [Co$_3$(η-C_5H_5)$_2$(μ-P(O)(OMe)$_2$)$_6$],[22] which is an important starting material for the syntheses of heterometallic di- and trinuclear phosphonatemetal complexes.[23,24]

D. (η⁵-CYCLOPENTADIENYL)BIS-(TRIMETHYLPHOSPHITE)RHODIUM(I)[25]

$$[Rh(P(OMe)_3)_2Cl]_2 + 2NaC_5H_5 \longrightarrow [Rh(\eta\text{-}C_5H_5)(P(OMe)_3)_2] + 2\,NaCl$$

Procedure

In a 125-mL Schlenk tube, equipped with a nitrogen inlet and a magnetic stirring bar, 1.0 g (2.03 mmol) of $[Rh(C_8H_{12})Cl]_2$ (ref.26) is dissolved in 50 mL of CH_2Cl_2. The solution is treated dropwise with 1 mL (8.1 mmol) of $P(OMe)_3$ and stirred for 2 hr at room temperature. The solvent and excess phosphite are removed *in vacuo*. The residue is dissolved in 60 mL of THF, and 450 mg (5.1 mmol) of NaC_5H_5 (ref. 20) is added. The reaction mixture is stirred for 24 hr at room temperature, filtered, and concentrated *in vacuo*. The oily residue is dissolved in 10 mL of hexane, and the solution is chromatographed on Al_2O_3 (neutral, activity III) using ether as eluant. The major yellow fraction is collected, the solvent is removed, and the resultant oily residue is dried for 24 hr in high vacuum ($\sim 10^{-4}$ torr). After storing the product for 3–4 weeks under an inert atmosphere without additional solvent at $-10°$, yellow crystals are formed. Yield: 605 mg (36%), mp 31°.

Anal. Calcd. for $C_{11}H_{23}O_6P_2Rh$: C, 31.74; H, 5.57; P, 14.88; MW, 416.15. Found: C, 31.84; H, 5.73; P, 14.72; MW, 416 (mass spectrometry).

Properties

(η-Cyclopentadienyl)bis(trimethylphosphite)rhodium(I) is a yellow crystalline low-melting product that is air sensitive and should be stored under an inert atmosphere at 10–20°. It is soluble in hydrocarbon solvents but decomposes in chloroform and carbon tetrachloride. The ¹H NMR spectrum (60 MHz, benzene-d_6, δ in ppm downfield from TMS) shows two signals at 5.32 (doublet of triplets, $J_{RhH} = 0.6$ Hz, $J_{PH} = 1.2$ Hz, C_5H_5 protons) and 3.40 [virtual triplet, $N = 12.2$ Hz, $P(OMe)_3$ protons].

The complex is a good nucleophile and can be used as starting material for the syntheses of cationic complexes $[Rh(\eta\text{-}C_5H_5)(P(OMe)_3)_2R]^+$ (R = H, Me).[21,27] It also reacts smoothly with alkali metal iodides MI (M = Li, Na, K) in two stages, via the intermediate $[Rh(\eta\text{-}C_5H_5)CH_3(P(OMe)_3)\text{-}P(O)(OMe)_2]$, to the corresponding rhodiumbis(phosphonate) complexes $[Rh(\eta\text{-}C_5H_5)CH_3(P(O)(OMe)_2)_2]M$.[28]

References

1. H. Werner, *Angew. Chem. Int. Ed. (Engl.),* **22,** 927 (1983).
2. H. Werner, R. Feser, and W. Buchner, *Chem. Ber.,* **112,** 834 (1979).
3. A. van der Ent and L. Onderdelinden, *Inorg. Synth.,* **14,** 92 (1973).
4. M. A. Lyle and S. R. Stobart, *Inorg. Synth.,* **17,** 178 (1977).
5. H. Werner, L. Hofmann, R. Feser, and W. Paul, *J. Organomet. Chem.,* **281,** 317 (1985).
6. J. M. Mayer and J. C. Calabrese, *Organometallics,* **3,** 1292 (1984).
7. H. Otto, Ph.D. Thesis, Universität Würzburg, 1986, p. 239.
8. H. F. Klein and H. H. Karsch, *Chem. Ber.,* **108,** 944 (1975).
9. H. Werner and W. Hofmann, *Chem. Ber.,* **110,** 3481 (1977).
10. R. T. Markham, E. A. Dietz Jr., and D. R. Martin, *Inorg. Synth.,* **16,** 153 (1976).
11. K. Dey and H. Werner, *Chem. Ber.,* **112,** 823 (1979).
12. H. Werner and W. Hofmann, *Chem. Ber.,* **114,** 2681 (1981).
13. H. Werner, K. Leonhard, and C. Burschka, *J. Organomet. Chem.,* **160,** 291 (1978).
14. O. Kolb and H. Werner, *J. Organomet. Chem.,* **268,** 49 (1984).
15. H. Werner, S. Lotz, and B. Heiser, *J. Organomet. Chem.,* **209,** 197 (1981).
16. C. Burschka, K. Leonhard, and H. Werner, *Z. Anorg. Allg. Chem.,* **464,** 30 (1980).
17. K. Leonhard and H. Werner, *Angew. Chem. Int. Ed. (Engl.),* **16,** 649 (1977).
18. H. Werner, K. Leonhard, O. Kolb, E. Röttinger, and H. Vahrenkamp, *Chem. Ber.,* **113,** 1654 (1980).
19. V. Harder, J. Müller, and H. Werner, *Helv. Chim. Acta,* **54,** 1 (1971).
20. R. B. King and F. G. A. Stone, *Inorg. Synth.,* **7,** 99 (1963).
21. H. Werner, H. Neukomm, and W. Kläui, *Helv. Chim. Acta,* **60,** 326 (1977).
22. V. Harder, E. Dubler, and H. Werner, *J. Organomet. Chem.,* **71,** 427 (1974).
23. W. Kläui and H. Werner, *Angew. Chem. Int. Ed. (Engl.),* **15,** 172 (1976).
24. W. Kläui and K. Dehnicke, *Chem. Ber.,* **111,** 451 (1978).
25. H. Neukomm and H. Werner, *Helv. Chim. Acta,* **57,** 1067 (1974).
26. J. Chatt and L. M. Venanzi, *J. Chem. Soc.,* **1957,** 4735.
27. H. Neukomm and H. Werner, *J. Organomet. Chem.,* **108,** C 26 (1976).
28. H. Werner and R. Feser, *Z. Anorg. Allg. Chem.,* **458,** 301 (1979).

39. MIXED COBALT–RUTHENIUM DODECACARBONYL CLUSTERS: DODECACARBONYLHYDRIDOTRI-COBALTRUTHENIUM, $Co_3RuH(CO)_{12}$

Submitted by F. OLDANI* and G. BOR†
Checked by T. J. COFFY‡ and S. G. SHORE‡

Reported yields of the early methods for the preparation of the mixed cobalt–ruthenium cluster hydride $Co_3RuH(CO)_{12}$ were generally very low

*Swiss Federal Institute of Technology, ETH, Zurich; present address: Ciba-Geigy Schweizerhalle AG, CH-8133 Schweizerhalle (BL), Switzerland.

†Swiss Federal Institute of Technology, ETH, High Pressure Laboratory, CH-8092 Zurich, Switzerland.

‡Department of Chemistry, Ohio State University, Columbus, OH 43210-1173.

$(\sim 7\%)$.[1,2] More recently yields improved up to $\sim 30\%$ have been reported.[3] In the first two methods[1,2] the ruthenium sources were $Ru_3(CO)_{12}$ and $[Ru(CO)_3Cl_2]_2$, respectively, whereas in the last method ruthenium was introduced in the form of $RuCl_3 \cdot 3H_2O$. We report here that the use of $Ru(CO)_5$ as a ruthenium source [which is easily obtained either thermally[4a] or photochemically[4b] from $Ru_3(CO)_{12}$] dramatically increases the overall yield in the synthesis of $Co_3RuH(CO)_{12}$ to 60–70%.

All these methods, including ours, yield the cluster hydride from the anionic intermediate $[Co_3Ru(CO)_{12}]^-$, by the addition of an acid in a polar solvent. We could also demonstrate in high pressure equipment that $Co_3RuH(CO)_{12}$ is directly formed in an apolar solvent from $Co_2(CO)_8$ and $Ru_3(CO)_{12}$ at 70° with a gas mixture composed of 100 bar H_2 and 0.6 bar CO. Under these conditions a mixture is gradually obtained that contains all $Co_{4-x}Ru_xH_x(CO)_{12}$ species ($x = 0, 1, 2, 3, 4$) in addition to $CoH(CO)_4$ and $Co_2(CO)_8$ in a stable equilibrium.[5,6] About 30% of the dissolved cobalt that is present in the form of different carbonyls is in $Co_3RuH(CO)_{12}$. This is in sharp contrast to the analogous iron–cobalt cluster, $Co_3FeH(CO)_{12}$,[7] which is thermodynamically unstable and is not formed from the components, but rather, when prepared through the anion $[Co_3Fe(CO)_{12}]^-$ decomposes in solution completely to yield $Fe(CO)_5$ and mixture of cobalt carbonyls.[5,6]

■ **Caution.** *All operations should be carried out in an efficient fume hood, since both escaping carbon monoxide and volatile metal carbonyls are toxic. Contact of the metal carbonyl solutions with skin and inhalation of their vapors should be avoided. Octacarbonyl dicobalt is oxidized readily by air and must be handled under an inert atmosphere.*

Procedure

$$Ru_3(CO)_{12} + 3CO \longrightarrow 3Ru(CO)_5 \tag{1}$$

$$\tfrac{7}{2} Co_2(CO)_8 + 2Ru(CO)_5 \longrightarrow [Co(solv)_x][Co_3Ru(CO)_{12}]_2 + 14CO \tag{2}$$

$$[Co(solv)_x][Co_3Ru(CO)_{12}]_2 + 2HCl \longrightarrow 2Co_3RuH(CO)_{12} + CoCl_2 \tag{3}$$

1. Dodecacarbonyltriruthenium (2.5 g, 3.91 mmol) prepared as reported,[8] is introduced into a 500-mL stainless steel (18-8-2) autoclave. Acetone (300 mL), purified by distillation over $CaCl_2$ or $CaSO_4 \cdot \tfrac{1}{2}H_2O$ (Sicon), is then added by suction. The autoclave is pressurized with carbon monoxide (220 bar), heated slowly to 154°, and then shuttled or stirred under these conditions for 22 hr. The maximum working pressure reached during this procedure was 435 bar.

The autoclave is then gradually cooled, first to room temperature, and then to $-78°$ in an acetone–dry ice mixture. Pressure is then released through a cold trap. The IR spectrum shows only the two C—O stretching bands of $Ru(CO)_5$ at 2042 and 1996 cm^{-1} (observed values in acetone solution). The yield is practically quantitative. This solution is used directly in the following steps. Note: In other solvents, like hexane or diethyl ether, the reaction is identical to that in acetone; the time required to the completion of the reaction can vary considerably, however.

2. Octacarbonyldicobalt (7.5 g, 22 mmol), prepared as reported,[9] is dissolved in purified acetone (100 mL) and stirred under CO in a 750-mL three-necked flask at 50° for 15 min. Then a previously prepared solution (300 mL) of $Ru(CO)_5$ [~12 mmol, cf. point (1)], cooled to $-78°$, is added in an atmosphere of CO. The combined solution is stirred at 60° for 17 hr. At the end of this period the solvent is removed at 60° *in vacuo*. The dry residue is kept at the same temperature under high vacuum ($<10^{-3}$ torr) for another 3–4 hr.

3. The dry product of part (2) is extracted in the same vessel by 10 portions of 100 mL each of distilled and deairated water under nitrogen. The resultant aqueous red solution is filtered under nitrogen and then acidified by the addition of hydrochloric acid (150 mL, 37%). The reddish-brown precipitate that is obtained in this way is filtered, washed with distilled water (20 mL), and dried at room temperature under high vacuum ($<10^{-3}$ torr) for 3 days. The red raw product, $Co_3RuH(CO)_{12}$ (4.8 g, 77 mmol, yield 61%) is recrystallized from a toluene–heptane mixture (1:1). The recrystallized sample and the raw product show the same IR spectrum.

Anal. calcd. for $C_{12}HO_{12}Co_3Ru$: C, 22.44; H, 0.16; Co/Ru, 3.0. Found: C, 22.64; H, 0.39; Co/Ru, 2.9.

Properties of $Co_3RuH(CO)_{12}$

$Co_3RuH(CO)_{12}$ is an air-stable, red, crystalline solid. It is soluble in non-polar solvents such as hexane, and very soluble in more polar or polarizable organic solvents such as benzene or dichloromethane. In THF (5% water) $Co_3RuH(CO)_{12}$ behaves as a strong acid as shown by potentiometric titration with piperidine.[5]

Solutions of $Co_3RuH(CO)_{12}$ in various hydrocarbons are stable in air for about 1 day.

The compound is best characterized by its IR spectrum in the C—O stretching region, which shows the following carbonyl bands (cm^{-1}) in hexane solution: 2063 (s), 2056 (s), 2022 (m), 2010 (w,sh), and 1886 (m). The IR spectrum of $Co_3RuH(CO)_{12}$ in acetone shows bands (cm^{-1}) at 2060 (w), 2017 (s), 2002 (s), 1970 (w), 1823 (w), which indicates the loss of the proton to give the cluster anion $[Co_3Ru(CO)_{12}]^-$. The mass spectrum of $Co_3RuH(CO)_{12}$ shows the parent ion at m/e 616 and ions resulting from the successive loss of 12 carbonyls. In the UV–vis region the heteronuclear cluster $Co_3RuH(CO)_{12}$ shows bands at 525 nm (3300 L·mol^{-1} cm^{-1}), 395 (5600) and 325 (10,000) (hexane solution). The C_{3v} symmetry of the molecule of $Co_3RuH(CO)_{12}$ is indicated by IR studies of the ^{13}CO-enriched compound.[5]

$Co_3RuH(CO)_{12}$ and/or salt of $[Co_3Ru(CO)_{12}]^-$ were reported to act as homogeneous catalysts in the homologation of methanol.[10,11] $Co_3RuH(CO)_{12}$ has been also applied to yield, by controlled thermal decomposition, very active heterogeneous Fischer–Tropsch catalysts.[12]

References

1. M. J. Mays and R. N. F. Simpson, *J. Chem. Soc. (A)*, **1968**, 1444.
2. D. B. W. Yawney and F. G. A. Stone, *J. Chem. Soc. (A)*, **1969**, 502.
3. M. Hidai, M. Orisaku, M. Ue, Y. Koyasu, T. Kodama, and Y. Uchida, *Organometallics*, **2**, 292 (1983).
4. (a) R. Whyman, *J. Organomet. Chem.*, **56**, 339 (1973);
   (b) B. F. G. Johnson, J. Lewis, and M. V. Twigg, *ibid.*, **67**, C75 (1974).
5. F. Oldani, Ph.D. thesis, No. 7476, ETH-Zurich, 1984.
6. G. Bor, *Pure Appl. Chem.*, **58**, 543 (1986).
7. P. Chini, L. Colli, and M. Peraldo, *Gazz. Chim. Ital.*, **90**, 1005 (1960).
8. B. R. James, G. L. Rempel, and W. K. Teo, *Inorg. Synth.*, **16**, 45 (1976); A. Mantovani and S. Cenini, *ibid.*, **16**, 47 (1976).
9. P. Szabó, L. Markó, and G. Bor, *Chem. Tech.* [*Leipzig*], **13**, 549 (1961). This method is more efficient than the previously reported ones in *Inorganic Synthesis:* compare *Inorg. Synth.*, **2**, 238 (1946) and **5**, 190 (1957).
10. M. Hidai, M. Orisaku, M. Ue, and Y. Uchida, *Chem. Lett.*, **1981**, 143; M. Hidai, M. Orisaku, M. Ue, Y. Koyasu, T. Kodama, and Y. Uchida, *Organometallics*, **2**, 292 (1983).
11. G. Doyle (Exxon Res. and Engng. Company), European Patent, 0030434/A1 (1981).
12. R. Hemmerich, W. Keim, and M. Röper, *J. Chem. Soc., Chem. Commun.*, **1983**, 428.

40. DINUCLEAR PHOSPHIDO AND ARSENIDO DERIVATIVES OF MOLYBDENUM

$[Mo(\eta^5\text{-}C_5H_5)(CO)_3]_2 + R_2EH \xrightarrow{\Delta}$

$\qquad\qquad [Mo(\eta^5\text{-}C_5H_5)(CO)_2]_2(\mu\text{-}H)(\mu\text{-}ER_2) + 2CO \uparrow$

A. E = P, R = *t*-Bu

B. E = As, R = Me

Submitted by R.A. JONES,* S.T. SCHWAB,* and A.L. STUART*
checked by J.L. PETERSEN†

A variety of dinuclear phosphido‡ or arsenido bridged complexes of molybdenum have been reported although their reactions have been little studied. The direct reaction of Me_2PH with $[Mo(\eta^5-C_5H_5)(CO)_3]_2$ gives $[Mo(\eta^5-C_5H_5)(CO)_2]_2(\mu-H)(\mu-Me_2P)$,[1] and several structural studies of this complex have been reported.[2-4] The preparations reported here are for the $t-Bu_2P$ and Me_2As analogs of this complex.[5] They may be prepared conveniently and in high yields, and they should prove to be useful starting materials for further investigation of the chemistry of phosphido or arsenido bridged complexes.

General Procedure

All reactions and operations must be carried out under dinitrogen or under vacuum, using standard Schlenk line techniques.[6] Diethyl ether, THF, and hexane are freshly distilled from sodium benzophenone ketyl under nitrogen. Toluene is freshly distilled from sodium under nitrogen. The compounds $[Mo(\eta^5-C_5H_5)(CO)_3]_2$,[1] $t-Bu_2PH$,[7] and Me_2AsH[8] are prepared as described in the literature cited.

A.   TETRACARBONYLBIS(η^5-CYCLOPENTADIENYL)-
μ-(DI-*tert*-BUTYLPHOSPHIDO)-
μ-HYDRIDODIMOLYBDENUM(2+)

Procedure

■ **Caution.** *Because of the toxicity of carbon monoxide and* t-Bu_2PH *this reaction should be carried out in an efficient fume hood.*

The compound hexacarbonylbis(η^5-cyclopentadienyl)dimolybdenum [Mo-$(\eta^5-C_5H_5)(CO)_3]_2$ (1.0 g, 2.04 mmol) is placed in a 250-mL Schlenk tube and 95 mL of toluene is added. Di-*tert*-butylphosphine ($t-Bu_2PH$) (0.596 g, 4.08 mmol) is added either via vacuum transfer or by syringe and the mixture is heated at reflux under nitrogen for 8 hr. During this time the color of the solution changes from red to brown. Evolved carbon monoxide is vented via a mercury bubbler. After cooling to room temperature, volatile materials are removed under vacuum, and the residue is extracted into two 15-mL portions of diethyl ether. The combined extracts are fil-

*Department of Chemistry, The University of Texas at Austin, Austin, TX 78712.
†Department of Chemistry, West Virginia University, Morgantown, WV 26506.
‡Also known as phosphino.

tered, and the solution is cooled to $-20°$. Red prisms of the product are formed in 72 hr. The supernatant liquid is decanted and the product dried under vacuum. Yield: 0.53 g (45%).

Anal. Calcd. for $C_{22}H_{29}Mo_2O_4P$: C, 45.53%; H, 5.04%. Found: C, 45.43%; H, 4.89%.

Properties[5]

The compound $[Mo(\eta^5-C_5H_5)(CO)_2]_2(\mu-H)(\mu-t-Bu_2P)$ is a red, crystalline, air-stable solid, mp 177 to 185° (dec). Infrared absorptions (CaF_2 cells, toluene) 1935 (s), 1857 (s), 1800 (w) cm^{-1}. Proton NMR (C_6D_6, 200 MHz) δ 4.95 (s) (10H), δ 1.28 (d) (J_{P-H} = 13.5 Hz, 18H), δ -13.1 (d), $^2J_{P-H}$ = 35.4 Hz) (in ppm rel. Me_4Si, δ 0.0). ^{31}P {^1H} NMR (C_6D_6, ambient temperature) δ 267.53 (s) (in ppm rel. 85% H_3PO_4, δ 0.0 at 32.384 MHz).

B. TETRACARBONYLBIS(η^5-CYCLOPENTADIENYL)-μ-(DIMETHYLARSIDO)-μ-HYDRIDODIMOLYBDENUM(2+)

Procedure

■ **Caution.** *Because of the toxicity of carbon monoxide and Me_2AsH (also pyrophoric) this reaction should be carried out in an efficient fume hood.*

The compound hexacarbonylbis(η^5-cyclopentadienyl)dimolybdenum $[Mo(\eta^5-C_5H_5)(CO)_3]_2$ (1.25 g, 2.55 mmol) is dissolved in 100 mL of THF in a 250-mL Schlenk tube. Dimethylarsine (Me_2AsH) (0.45 mL, 5.14 mmol) is added either via vacuum transfer or by syringe, and the solution is heated under reflux for 40 hr. The resulting deep red solution is filtered, and volatile materials are removed under vacuum. The residue is washed with two 15-mL portions of hexane, and then extracted into two 15-mL portions of diethyl ether. The combined extracts are filtered, and the volume of the solution is reduced to 15 mL by evaporation under vacuum. Cooling to $-20°$ gives the product as orange, air-stable rods. A small amount ($\sim 10\%$) of another product $(\eta^5-C_5H_5)Mo(CO)_2(\mu-AsMe_2)(\mu-As_2Me_4)Mo(CO)_4$ also crystallizes as yellow prisms. Fractional crystallization of the first crop of crystals can be used to seperate the two compounds since compound B is more soluble and crystallizes after $(\eta^5-C_5H_5)Mo(CO)_2(\mu-AsMe_2)(\mu-As_2Me_4)Mo(CO)_4$. Yield of B: 0.65 g (50%), mp 185 to 192° (dec).

Anal. Calcd. for $C_{16}H_{17}AsMo_2O_4$: C, 35.58%; H, 3.17%. Found: C, 35.62%; H, 3.10%.

Properties[5]

The product is moderately soluble in most polar organic solvents. The IR spectrum (THF solution, CaF_2 cells) shows two strong bands at 1935 and 1870 cm^{-1}. The ^1H NMR spectrum has peaks at δ 4.65 (s) (10H), δ 1.54 (s) (6H) and δ -12.09 (s) (1H) (in C_6D_6, at ambient temperature, 90 MHz, ppm rel. Me_4Si, δ 0.0).

References

1. R. G. Hayter, *Inorg. Chem.*, **2**, 1031 (1963).
2. R. J. Doedens and L. F. Dahl, *J. Am. Chem. Soc.*, **87**, 2576 (1965).
3. J. L. Petersen, L. F. Dahl, and J. M. Williams, *J. Am. Chem. Soc.*, **96**, 6610 (1974).
4. J. L. Petersen and J. M. Williams, *Inorg. Chem.*, **17**, 1308 (1978).
5. R. A. Jones, S. T. Schwab, A. L. Stuart, B. R. Whittlesey, and T. C. Wright, *Polyhedron*, **4**, 1689 (1985).
6. D. F. Shriver, *Manipulation of Air-Sensitive Compounds*, McGraw-Hill, New York, 1969.
7. H. Hoffman and P. Schellenbeck, *Chem. Ber.*, **99**, 1134 (1966); K. Issleib and F. Krech, *J. Organomet. Chem.*, **13**, 283 (1968).
8. R. D. Feltham and W. Silverthorn, *Inorg. Synth.*, **10**, 159 (1967).

41.   DICARBONYLBIS(DI-*tert*-BUTYLPHOSPHINE)- (μ-DI-*tert*-BUTYLPHOSPHIDO)-μ-HYDRIDO- DIRHODIUM(1 +)(Rh_2(μ-*t*-Bu_2P)(μ-H)(CO)$_2$(*t*-Bu_2PH)$_2$ AND μ-CHLORO-BIS(η^4-1,5-CYCLOOCTADIENE)(μ-DI-*tert*- BUTYLPHOSPHIDO)DIRHODIUM(1 +) (Rh_2(μ-*t*-Bu_2P)- μ-Cl(COD)$_2$),(COD = 1,5-CYCLOOCTADIENE)

$$Rh_4(CO)_{12} + 6 \cdot t\text{-}Bu_2PH \xrightarrow{\text{toluene}}$$

$$Rh_2(\mu\text{-}t\text{-}Bu_2P)(\mu\text{-}H)(CO)_2(t\text{-}Bu_2PH)_2 + 8CO \uparrow$$
$$\textbf{A}$$

$$[Rh(COD)Cl]_2 + Li\text{-}t\text{-}Bu_2P \xrightarrow{\text{hexane}}$$

$$Rh_2(\mu\text{-}t\text{-}Bu_2P)(\mu\text{-}Cl)(\eta^4\text{-}COD)_2 + LiCl \downarrow$$
$$\textbf{B}$$

Submitted by R. A. JONES,* D. E. HEATON,* and T. C. WRIGHT*
Checked by D. W. MEEK† and C. SUPPLEE†

Di-*tert*-butylphosphido‡ (*t*-Bu_2P) complexes of the *d*-block transition metals have been synthesized in order to study the steric effects of phosphido

*Department of Chemistry, The University of Texas at Austin, Austin, TX 78712.
†Department of Chemistry, The Ohio State University, Columbus, OH 43210-1173.
‡Also known as phosphino.

ligands in organometallic chemistry.[1-3] Most of the syntheses so far reported are tedious, difficult, or give low yields. The syntheses reported here are straightforward and give reasonable yields of dinuclear Rh(I) species with one *t*-Bu$_2$P bridging ligand. They should prove to be useful starting materials for further studies in this area.

General Procedure

Tetrahydrofuran (THF) and hexane are freshly distilled from sodium benzophenone ketyl under nitrogen. Toluene is freshly distilled from sodium metal under nitrogen. The starting materials; Rh$_4$(CO)$_{12}$,[4] [Rh(COD)Cl]$_2$,[5] *t*-Bu$_2$PH, and *t*-Bu$_2$PLi (in THF)[6] are prepared as previously described. All manipulations are performed under nitrogen or under vacuum, using standard procedures.[7]

A.   DICARBONYLBIS(DI-*tert*-BUTYLPHOSPHINE)(μ-DI-*tert*-BUTYLPHOSPHIDO)-μ-HYDRIDO-DIRHODIUM(1 +) [Rh$_2$(μ-*t*-Bu$_2$P)(μ-H)(CO)$_2$(*t*-Bu$_2$PH)$_2$]

Procedure

■ **Caution.** *Because of the high toxicity of carbon monoxide this reaction should be carried out in an efficient fume hood.*

Dodecacarbonyltetrarhodium [Rh$_4$(CO)$_{12}$] (0.20 g, 0.27 mmol) is dissolved in 50 mL of toluene in a 100-mL Schlenk flask. To the red solution is added di-*tert*-butylphosphine (*t*-Bu$_2$PH) (0.22 mL, 1.62 mmol) via syringe.* A reflux condenser is attached to the flask and the orange solution is stirred magnetically and heated under reflux for 12 hr. During this time the solution becomes dark red in color. Evolved carbon monoxide (■ **CAUTION**, *very toxic*) is vented off via a mercury bubbler. After cooling to room temperature volatile materials are removed under vacuum (oil vacuum pump, ~10^{-2} torr). The residue is extracted with two 25-mL portions of hexane, which are then combined and filtered. The volume of the solution is reduced to ~10 mL under vacuum and cooling to −40° overnight gives large yellow prisms of Rh$_2$(μ-*t*-Bu$_2$P)(μ-H)(CO)$_2$(*t*-Bu$_2$PH)$_2$, which are removed by decantation from the supernatant liquid and dried under vacuum. Yield: 0.23 g (61% based on Rh).

*The stoichiometry is critical if Rh:P = 6:4 is used the cluster Rh$_6$ (μ − *t* − Bu$_2$P)$_4$ (co)$_8$ (μ − H)$_2$ is produced.

Anal. Calcd. for $C_{26}H_{57}O_2P_3Rh_2$: C, 44.57%; H, 8.14%; P, 13.29%. Found: C, 44.33%; H, 8.05%; P, 12.89%.

Properties

Dicarbonylbis(di-*tert*-butylphosphine)(μ-di-tert-butylphosphido)-μ-hydri-dodirhodium(1+) is a yellow crystalline material that loses crystallinity under vacuum. It is soluble in common organic solvents such as hexane or toluene. It is moderately air sensitive in the solid state and more so when in solution.

The ^{1}H NMR spectrum, in toluene-d_8 solution at $-80°$, shows resonances at δ 4.70 (d, $J_{P-H} = 308.7$ Hz) (2H, *t*-Bu$_2$P*H*); δ 1.63 ($J_{P-H} = 6.69$ Hz) (18H, μ-*t*-Bu$_2$P); δ 1.18 (d, $J_{P-H} = 11.69$ Hz) (36H, *t*-Bu$_2$PH); δ $-$10.40 (*m*) (1H, μ-H) (in ppm rel. ext. *Me*$_4$Si; δ 0.0 at 90 MHz). The ^{31}P $\{^1$H$\}$ NMR in THF at ambient temperature shows a triplet of triplets (t.t.) at δ 264.3 (μ-*t*-Bu$_2$P) ($^1J_{Rh-P} = 113$ Hz, $^2J_{P-P} = 190$ Hz) and a doublet of doublets (d.d.) at δ 58.5, $^1J_{Rh-P} = 123$ Hz, $^2J_{P-P} = 190$ Hz (*t*-Bu$_2$P*H*) (at 32.384 MHz in ppm rel. 85% H$_3$PO$_4$(aq), δ 0.0),* mp 142–148° (dec), IR (Nujol mull, KBr plates) 2292 (m), 1942 (vs), 1903 (sh) cm^{-1}. The solid state structure has been determined by a single crystal X-ray diffraction study. The coordination geometry about each Rh atom is essentially planar with a fairly long Rh—Rh single bond of 2.905 Å.[8]

B. μ-CHLORO-BIS(η^4-1,5-CYCLOOCTADIENE)-(μ-DI-*tert*-BUTYLPHOSPHIDO)-DIRHODIUM(1+) [Rh$_2$(μ-*t*-Bu$_2$P)μ-Cl(COD)$_2$]

Procedure

In a 250-mL Schlenk flask is placed [Rh(η^4-COD)Cl]$_2$ (0.63 g, 1.27 mmol) and 80 mL of hexane. The yellow suspension is stirred magnetically and cooled to 0°. A solution of *t*-Bu$_2$PLi (2.0 mL of a 0.66 *M* THF solution, 1.32 mmol) is added dropwise via syringe. The red-orange mixture is then allowed to warm to room temperature over ~0.5 hr. The flask is then fitted with a reflux condenser and the mixture is heated under reflux for ~16 hr until the suspended [Rh(η^4-COD)Cl]$_2$ is dissolved. The dark orange solution is then cooled to room temperature and volatile materials are removed under vacuum. The residue is extracted with three 50-mL portions of toluene and the extracts are combined and filtered. The solution is concen-

---

*The checkers found ^{31}P $\{^1$H$\}$ NMR at 202.5 MHz, δ = 259.1, (t.t., $^1J_{Rh-P} = 113$ Hz, $^2J_{P-P} = 190$ Hz), δ = 54.5 (d.d., $^1J_{Rh-P} = 121$ Hz, $^2J_{P-P} = 190$ Hz) for A.

trated under vacuum to ~20 mL and cooled to − 40° to yield yellow crystals of $Rh_2(\mu\text{-}t\text{-}Bu_2P)(\mu\text{-}Cl)(\eta^4\text{-}COD)_2$. Yield: 0.42 g (55% based on Rh).

Anal. Calcd. for $C_{24}H_{42}ClPRh_2$: C, 47.80%; H, 6.97%; P, 5.15%. Found: C, 47.45%; H, 6.53%; P, 4.92%.

Properties

μ-Chloro-bis(η^4-1,5-cyclooctadiene)(μ-di-*tert*-butylphosphino)-dirhodium(1+) is a yellow, crystalline, moderately air-stable, hexane-soluble complex, mp 185–190° (dec). The ^{31}P {^1H} NMR spectrum in benzene-d_6 at ambient temperature shows a triplet at δ 43.2 ppm ($^1J_{Rh-P} = 114$ Hz).[8] The crystal structure of this complex contains two planar Rh(I) units with no apparent direct bonding interactions between the metal centers (Rh—Rh = ~3.3 Å).[8]

References

1. R. A. Jones, A. L. Stuart, J. L. Atwood, and W. E. Hunter, *Organometallics*, **2**, 1437 (1983).
2. R. A. Jones, T. C. Wright, J. L. Atwood, and W. E. Hunter, *Organometallics*, **2**, 470 (1983).
3. R. A. Jones and T. C. Wright, *Organometallics*, **2**, 1842 (1983).
4. P. E. Cattermole and A. G. Osborne, *Inorg. Synth.*, **17**, 115 (1977).
5. G. Giordano and R. H. Crabtree, *Inorg. Synth.*, **19**, 218 (1979).
6. H. Hoffman and P. Schellenbeck, *Chem. Ber.*, **99**, 1134 (1966); K. Issleib and F. Krech, *J. Organomet. Chem.*, **13**, 283 (1968).
7. D. F. Shriver, *Manipulation of Air-Sensitive Compounds*, McGraw-Hill, New York, 1969.
8. A. M. Arif, R. A. Jones, M. H. Seeberger, B. R. Whittlesey, and T. C. Wright, *Inorg. Chem.*, **25**, 3943 (1986).

42. *tert*-BUTYLPHOSPHIDO (t-BuP(H)$^-$) BRIDGED DIMERS OF RHODIUM(+1) AND NICKEL(+1) CONTAINING Rh=Rh DOUBLE AND Ni—Ni SINGLE BONDS

$$[Rh(\eta^4\text{-}COD)Cl]_2 + 2(H)(t\text{-}Bu)PLi \xrightarrow[PMe_3]{Et_2O}$$

+ LiCl + other products   (A)

COD = 1,5-cyclooctadiene

$$NiCl_2(PMe_3)_2 + 2(H)(t\text{-Bu})PLi \xrightarrow{Et_2O}$$

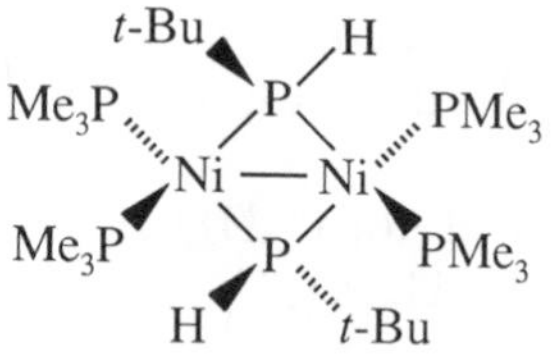

(B)

Submitted by R. A. JONES* and M. H. SEEBERGER*
Checked by A. A. CHERKAS,† F. VAN GASTEL,† and A. J. CARTY†

Relatively few phosphido complexes of the transition metals are known that have an alkyl or aryl group (R) in addition to hydrogen attached to phosphorus [i.e., R(H)P⁻].[1] The reactivity of the P—H unit has been studied in only a few cases.[2] The syntheses described here are for dinuclear Rh(+1) or Ni(+1) complexes that have two *t*-BuP(H)⁻ bridges. In both complexes the metals have pseudotetrahedral geometries. Of added interest for the rhodium complex is the presence of a Rh=Rh double bond.[3]

General Procedure

All operations are performed under nitrogen or under vacuum using standard Schlenk techniques.[4] Diethyl ether, hexane, and THF are freshly distilled from sodium benzophenone ketyl. The compounds $[Rh(\eta^4\text{-}COD)Cl]_2$,[5] $NiCl_2(PMe_3)_2$,[6] and $t\text{-BuPH}_2$[7] are prepared by the literature methods. A solution of the compound $(H)(t\text{-Bu})PLi$ is prepared by the addition of *n*-BuLi to a solution of $t\text{-BuPH}_2$ in THF (see below).

A. BIS-*tert*-BUTYLPHOSPHIDOTETRAKIS-(TRIMETHYLPHOSPHINE)DIRHODIUM(+1) [Rh(μ-*t*-Bu(H)P)(PMe₃)₂]₂

Procedure

$$t\text{-BuPH}_2 + n\text{-BuLi} \xrightarrow[-78°]{THF} (H)(t\text{-Bu})PLi + n\text{-BuH} \uparrow$$

A dilute (~0.5 *M*) solution of $(H)(t\text{-Bu})PLi$ in THF is prepared by the addition of one equivalent of *n*-BuLi to $t\text{-BuPH}_2$ in THF.

*Department of Chemistry, The University of Texas at Austin, Austin, TX 78712.
†Department of Chemistry, The University of Waterloo, Waterloo, Ontario, Canada N2L 3G1.

A 250-mL Schlenk tube, flushed with nitrogen and capped with a rubber septum, is cooled to −78° (dry ice/acetone bath) and THF (50 mL) is added. t-BuPH$_2$ (3.68 g, 5.0 mL, 41 mmol) is then added via syringe. Exactly one equivalent of n-BuLi (18.6 mL of a 2.20 M hexane solution) is then added slowly via syringe while the mixture is stirred magnetically. The evolved butane is vented via a mercury bubbler. The solution is allowed to warm to room temperature over 4 hr. The yellow-orange solution is then used for subsequent reactions. It may be stored at −40° for several weeks. If a slight excess of n-BuLi is used, a darker orange color results although this does not appear to affect the final yield of either compound formed by procedures A or B.

The compound di-μ-chloro-bis(η^4-1,5-cyclooctadiene)dirhodium,[5] (0.95 g, 1.93 mmol) is placed in a 250-mL Schlenk tube, and 100 mL of diethyl ether is added via syringe. The tube is cooled to −100° (liquid nitrogen, ethanol slush), and a solution of (H)(t-Bu)PLi in THF (6.57 mL of a 0.56 M solution, 3.68 mmol), prepared as just described, is added dropwise via syringe. The mixture is stirred magnetically and allowed to warm to room temperature over 10 hr. At this stage, an intensely red-purple solution is produced. The Schlenk tube is then recooled to −100° and excess PMe$_3$ **(CAUTION, TOXIC)** (2.0 mL, 10 mmol) is added via syringe. The mixture is again allowed to warm to room temperature (over 2 hr). (The checkers found a green-brown solution at this stage.) The solution is then filtered, and the volatile materials are removed under vacuum. The residue is extracted into 40 mL of hexane. (The checkers used 90 mL of hexane.) The solution is filtered and evaporated under vacuum to 10 mL. Cooling to −40° for 16 hr yields a large crystalline mass of the product. The supernatant liquid is decanted and the product is dried under vacuum at room temperature.

Anal. Calcd. for C$_{20}$H$_{56}$P$_6$Rh$_2$: C, 34.87%; H, 8.10%; P, 27.02%. Found: C, 35.49%; H, 8.38%; P, 26.38%.

Properties

The product is a red, crystalline, hexane-soluble material.[1] It is air stable in the solid state for short periods but unstable in solution when exposed to air. Yield: 85% (checkers found 74%), mp 219–221° (dec). The IR spectrum (Nujol mull, KBr plates) has a band of medium intensity at 2140 cm^{-1} assigned to the P—H stretch. The ^{31}P $\{^1$H$\}$ NMR at 32.384 MHz in C$_6$D$_6$ at ambient temperature is a second-order pattern with two multiplets centered at δ 207.25 and δ − 10.5 (in ppm rel. 85% H$_3$PO$_4$, δ-0.0).[1]

B. BIS-*tert*-BUTYLPHOSPHIDOTETRAKIS-(TRIMETHYLPHOSPHINE)DINICKEL(+1)
[Ni(μ-*t*-Bu(H)P)(PEe$_3$)$_2$]$_2$

Procedure

The compound dichlorobis(trimethylphosphine)nickel[6] (2.11 g, 7.4 mmol) is placed in a 250-mL Schlenk tube, and 80 mL of diethyl ether is added. The mixture is then cooled to $-100°$ (liquid nitrogen, ethanol slush), and a solution of (H)(*t*-Bu)PLi (26.4 mL of a 0.56 *M* THF solution, 14.8 mmol) is added dropwise via syringe. The reaction mixture is stirred magnetically and allowed to warm to room temperature over a period of 10 hr. The solution is then filtered, the volatile materials are removed under vacuum, and the residue is extracted into 80 mL of hexane. The solution is filtered and concentrated to 15 mL under vacuum. Cooling to $-40°$ gives a black solid after 12 hr. The supernatant liquid is decanted, and the solid is redissolved in hexane (40 mL). The resulting intensely purple colored solution is again concentrated (15 mL) and cooled ($-40°$). After 48 hr, large, deep purple crystals of the product are formed. They are separated from the supernatant liquid and dried under vacuum. Yield: 1.85 g (85%).

Anal. Calcd. for C$_{20}$H$_{56}$Ni$_2$P$_6$: C, 40.00%; H, 9.33%; P, 31.00%. Found: C, 39.37%; H, 8.54%; P, 30.38%.

Properties

The product is deep purple (almost black) in the solid state and gives intensely colored purple solutions in hydrocarbon solutions. It is best handled under nitrogen at all times since it decomposes within minutes on exposure to the atmosphere, mp 119–121° (dec slowly over this range), IR (Nujol mull, KBr plates) 2160 (m) (cm^{-1}) ν_{P-H}. The ^{31}P {^{1}H} NMR shows resonances at δ 160.00 (s), δ 136.54 (s), δ-22.13 (s), δ-32.00 (in C$_6$D$_6$ solution at 32.384 MHz, in ppm rel. 85% H$_3$PO$_4$ δ 0.0).[1] (The checkers found IR; $\nu_{P-H} = 2173$ cm^{-1} and ^{31}P {^{1}H} NMR (101.26 MHz, toluene-d_8; δ-224.4 (s), 138.9 (s), 32.9 (s), -19.9 (s), -21.2 (s), -21.5 (s), (the unusual ^{31}P NMR spectrum has been discussed previously[1]) mp 119–121°).

References

1. R. A. Jones, N. C. Norman, M. H. Seeberger, J. L. Atwood, and W. E. Hunter, *Organometallics*, **2**, 1629 (1983).
2. H. Vahrenkamp and D. Wolters, *Angew. Chem. Int. Ed. (Engl.)*, **22**, 154 (1983); H. Vahrenkamp and D. Wolters, *J. Organomet. Chem.*, **224**, C17 (1982); P. M. Treichel,

W. K. Dean, and W. M. Douglas, *Inorg. Chem.*, **11**, 1609 (1972); W. Clegg and S. Morton, *Inorg. Chem.*, **18**, 1189 (1979); J. S. Field, R. J. Haines, and D. Smith, *J. Organomet. Chem.*, **224**, C49 (1982); K. Isslieb and H. R. Roloff, *Z. Anorg. Allg. Chem.*, **324**, 250 (1963).
3. R. A. Jones and T. C. Wright, *Organometallics*, **2**, 1842 (1983).
4. D. F. Shriver, *Manipulation of Air-Sensitive Compounds*, McGraw-Hill, New York, 1969.
5. G. Giordano and R. H. Crabtree, *Inorg. Synth.*, **19**, 218 (1979).
6. O. Dahl, *Acta Chem. Scand.*, **23**, 2342 (1969).
7. M. C. Hoff and P. Hill, *J. Org. Chem.*, **24**, 356 (1959).

43. TETRACARBONYLBIS(μ-DI-*tert*-BUTYLPHOSPHIDO)DICOBALT(+1) [Co(μ-*t*-Bu₂P)(CO)₂]₂

$$Co_2(CO)_8 + 2t\text{-}Bu_2PH \xrightarrow[\Delta]{\text{toluene}}$$

Submitted by D. J. CHANDLER,* R. A. JONES,* K. S. RATLIFF,*
and A. L. STUART*
Checked by N. VISWANATHAN† and G. L. GEOFFROY†

Varieties of dinuclear cobalt complexes are known with two phosphido bridges.[1] Those so far reported with the di-*tert*-butylphosphido ligand have two pseudotetrahedral 18-electron Co atoms linked by a metal–metal bond formally of order 2.[1] Tetracarbonylbis(μ-di-*tert*-butylphosphido)-dicobalt(+1) can be prepared from the reaction of Co(CO)₄I (generated *in situ*) with Li(*t*-Bu₂P) in THF.[1] We describe here a simplified, high yield synthesis of this dimer via the interaction of Co₂(CO)₈ with *t*-Bu₂PH in toluene. The complex should prove to be a useful starting material for further reactivity studies.

General Procedure

Hexane is freshly distilled from sodium benzophenone ketyl under nitrogen. Toluene is freshly distilled from sodium under nitrogen. Octacarbonyldicobalt {Co₂(CO)₈} (Strem Chemicals) is used as obtained. Di-*tert*-butylphosphine is prepared by the literature method.[2] All manipulations are

*Department of Chemistry, The University of Texas at Austin, Austin, TX 78712.
†Department of Chemistry, The Pennsylvania State University, University Park, PA 16802.

performed under nitrogen or under vacuum using standard Schlenk line techniques.[3]

Procedure

■ **Caution.** *Because of the high toxicity of carbon monoxide, this reaction should be carried out in an efficient fume hood.*

Octacarbonyldicobalt {$Co_2(CO)_8$} (4.00 g, 11.7 mmol) is placed in a 250-mL Schlenk tube, which is then cooled to $-78°$ with a dry ice–acetone bath. Toluene (150 mL) is added, and the reddish-brown suspension is stirred magnetically. Di-*tert*-butylphosphine (t-Bu_2PH) (3.42 g, 23.4 mmol) is added via syringe, and the mixture is allowed to warm to room temperature. It is first brown, becomes reddish brown and then burgundy in color. The solution is heated under reflux, under nitrogen for 12 hr, and the color changes first to dark red and then almost black. The carbon monoxide that is evolved is allowed to escape via a mercury bubbler (■ **CAUTION. VERY TOXIC**). After cooling to room temperature, volatile materials are removed under vacuum. The residue is washed with 20 mL of hexane and then extracted into four 40-mL portions of toluene. The dark red hexane solution is discarded, the extracts are combined, and the solution is filtered. Some dark green residue is left on the filter. The volume of the solution is reduced to ~100 mL under vacuum at room temperature and cooled to $-40°$. After 24 hr, dark green crystals of the product are removed from the supernatant liquid and dried under vacuum. Additional product can be obtained from the supernatant liquid by further reduction in volume and recooling. Overall yield: 4.55 g (75%), mp 215–218° (sealed tube under nitrogen, 1 atm, uncorrected) (dec).

Anal. Calcd. for $C_{20}H_{36}Co_2O_4P_2$: C, 46.1%; H, 6.92%; P, 11.9%. Found: C, 45.9%; H, 6.79%; P, 11.7%.

Properties

Tetracarbonylbis(μ-di-*tert*-butylphosphido)dicobalt(+1) is a deep green, almost black, crystalline solid. It is air-stable in the solid state for several hours. Solutions decompose slowly when exposed to air. The compound is soluble in toluene, benzene, and THF but only sparingly so in hexane. The IR spectrum (toluene solution, KBr cells) shows two strong bands at 1997 and 1955 cm^{-1}. The ^1H NMR spectrum in C_6D_6 at ambient temperature shows a multiplet at δ 1.15 (t-Bu_2P) (in ppm rel. Me_4Si δ 0.0,

90 MHz). The ^{31}P $\{^1H\}$ NMR spectrum shows a broad singlet; δ 331.43 ($\Delta w_{1/2}$ = 33 Hz) (in ppm rel. 85% H_3PO_4 (aq), δ 0.0).[1]

References

1. See, for example, R. A. Jones, A. L. Stuart, J. L. Atwood, and W. E. Hunter, *Organometallics*, **2**, 1437 (1983) and references therein.
2. H. Hoffman and P. Schellenbeck, *Chem. Ber.*, **99**, 1134 (1966); K. Issleib and F. Krech, *J. Organomet. Chem.*, **13**, 283 (1968).
3. D. F. Shriver, *Manipulation of Air-Sensitive Compounds*, McGraw-Hill, New York, 1969.

44. (η⁵-CYCLOPENTADIENYL)DIRUTHENIUM COMPLEXES

Submitted by N. M. DOHERTY* and S. A. R. KNOX*
Checked by C. P. CASEY† and G. T. WHITEKER†

The generation and interconversion of hydrocarbon fragments on metal surfaces is an important aspect of transition metal catalysis.[1] In an effort to model and understand these transformations, much attention has been focused on the synthesis and reactivity of organic species coordinated at polynuclear transition metal centers.[2] Organodiruthenium complexes have provided a particularly rich area of study. The availability of a variety of organometallic derivatives of the bis(η⁵-cyclopentadienyl)diruthenium carbonyl system has allowed extensive examination of the reactivity of bridging alkylidene, alkylidyne, and ethenylidene ligands.

The starting material for preparation of these derivatives, [Ru$_2$(CO)$_4$(η⁵-C$_5$H$_5$)$_2$], has previously been obtained by the reaction of sodium cyclopentadienide with a dihaloruthenium(II) carbonyl $\{$[Ru(CO)$_2$I$_2$] (ref. 3) or [Ru(CO)$_3$Cl$_2$]$_2$(ref. 4)$\}$ prepared by carbonylation of the corresponding ruthenium(III) trihalide. A more facile synthesis was later reported, involving the reaction of triruthenium dodecacarbonyl with cyclopentadiene.[5] The procedure described herein represents a modification of this second method, resulting in an improved yield.

A number of synthetic routes to organic derivatives of [Ru$_2$(CO)$_4$(η⁵-C$_5$H$_5$)$_2$] have been developed[6,7] and several representative examples are described here. One convenient entry into this chemistry is provided by the complex [Ru$_2$(CO)(μ-CO)$\{$μ-η^1:η^3-C(O)C$_2$Ph$_2\}$(η⁵-C$_5$H$_5$)$_2$](I).[8]

*Department of Inorganic Chemistry, The University, Bristol BS8 1TS, United Kingdom.
†Department of Chemistry, University of Wisconsin, Madison, WI 53706.

I

In boiling toluene, diphenylacetylene is rapidly displaced from this species by a variety of reagents; that is, it serves as an excellent source of the $Ru_2(CO)_3(\eta^5-C_5H_5)_2$ fragment.[9] Reaction with phosphorus ylides provides a series of μ-alkylidene derivatives $[Ru_2(CO)_2(\mu\text{-}CO)(\mu\text{-}CHR)(\eta^5\text{-}C_5H_5)_2]$.[10] Another route to organodiruthenium complexes involves anionic methyl attack on a coordinated carbonyl ligand of $[Ru_2(CO)_4(\eta^5\text{-}C_5H_5)_2]$, followed by protonation to generate a cationic μ-alkylidyne derivative, $[Ru_2(CO)_2(\mu\text{-}CO)(\mu\text{-}CMe)(\eta^5\text{-}C_5H_5)_2]^+$. This can be deprotonated to give a μ-ethenylidene complex $[Ru_2(CO)_2(\mu\text{-}CO)(\mu\text{-}CCH_2)(\eta^5\text{-}C_5H_5)_2]$ or treated with hydride to yield the μ-ethylidene derivative $[Ru_2(CO)_2(\mu\text{-}CO)(\mu\text{-}CHMe)(\eta^5\text{-}C_5H_5)_2]$.[11]

A. TETRACARBONYLBIS(η^5-CYCLOPENTADIENYL)-DIRUTHENIUM

$$\tfrac{2}{3}[Ru_3(CO)_{12}] + 2C_5H_6 \longrightarrow$$
$$2[RuH(CO)_2(\eta^5\text{-}C_5H_5)] \xrightarrow{O_2} [Ru_2(CO)_4(\eta^5\text{-}C_5H_5)_2]$$

Procedure

A three-necked, 500-mL, round-bottomed flask equipped with a nitrogen by-pass, a reflux condenser, and a magnetic stirring bar is charged with triruthenium dodecacarbonyl (8.5 g, 0.013 mol) (best prepared by the carbonylation of $RuCl_3$ in methanol[12]), 350 mL of dry, deoxygenated heptane, and freshly distilled cyclopentadiene (17.5 g, 0.265 mol) [prepared by cracking dicyclopentadiene over iron filings under a nitrogen atmosphere, and collecting the cyclopentadiene distillate (40–45°) from a 12-in. fractionating column]. The mixture is heated at reflux for 1 hr, producing $[RuH(CO)_2(\eta^5\text{-}C_5H_5)]$. A stopper is then removed from the flask and the volume of solvent is reduced to 50 mL by continued heating at reflux under a brisk flow of nitrogen, allowing the heptane to boil away. (This procedure is performed in a fume hood.) At this point orange product begins to crystallize from the reaction mixture. An additional 300 mL of untreated heptane, obtained directly from the reagent bottle, is added to the flask, the stopper is replaced, and the solution is heated at reflux for a further 2 hr. On cooling to room temperature the reaction mixture affords orange crystals of the product. The solid is collected by decantation and washed three times with 30-mL portions of hexane. After drying under vacuum the yield is 7.15 g. Additional product is obtained from the decanted solution and hexane washings by evaporation of the solvent followed by chromatography on a 3 × 20 cm alumina (Brockman grade 2) column. Elution with dichloromethane–hexane (1:3) removes unreacted starting materials and impurities as yellow bands. Elution with dichloromethane–hexane (1:1) develops a yellow band from which 0.77 g of product is obtained. Overall yield of product is 90–95% by this procedure.

Anal. Calcd. for $C_{14}H_{10}O_4Ru_2$: C, 37.8; H, 2.3. Found: C, 37.6; H, 2.1.

Properties

The compound is an air-stable, orange crystalline solid, soluble in common organic solvents. Solutions decompose slowly in air on exposure to light. The IR spectrum (in CH_2Cl_2) shows bands at 2003 (s), 1966 (s), 1934 (m), and 1771 (s) cm^{-1} due to the carbonyl ligands.

B. μ-CARBONYL-CARBONYLBIS(η⁵-CYCLOPENTADIENYL)-(μ-3-OXO-1,2-DIPHENYL-1-η:1,2,3-η-1-PROPEN-1,3-diyl)-DIRUTHENIUM(Ru—Ru), [Ru₂(CO)(μ-CO){μ-η¹:η³-C(O)C₂Ph₂}(η⁵-C₅H₅)₂] (I)

$$[Ru_2(CO)_4(\eta^5\text{-}C_5H_5)_2] + PhC_2Ph \xrightarrow{h\nu}$$
$$[Ru_2(CO)(\mu\text{-}CO)\{\mu\text{-}\eta^1\text{:}\eta^3\text{-}C(O)C_2Ph_2\}(\eta^5\text{-}C_5H_5)_2] + CO$$

Procedure

The reaction is performed in a silica glass reaction tube (2.5 cm in diameter, 40 cm in length, with a ground glass joint at the top) equipped with a nitrogen by-pass and a magnetic stirring bar. The UV irradiation source is a 250-W mercury lamp held ~20 cm from the reaction vessel. A mixture of tetracarbonylbis(η⁵-cyclopentadienyl)diruthenium (1.40 g, 3.15 mmol) and diphenylacetylene (1.68 g, 9.43 mmol) in dry, deoxygenated toluene (150 mL) is irradiated under a nitrogen atmosphere for 40 hr, during which time the color of the solution changes from orange to dark red. The solvent is evaporated from this mixture under reduced pressure and the residue is dissolved in the minimum of dichloromethane, then introduced to an alumina (Brockman grade 2) column (20 × 3 cm). Elution with dichloromethane–hexane (7:3) separates a yellow band that yields 0.045 g (3%) of a yellow crystalline side product $[Ru_2(CO)_4(\eta^5-C_5H_5)\{\eta^5-C_5H_4Ru(CO)_2(\eta^5-C_5H_5)\}]$.[8] Elution with dichloromethane–acetone (20:1) develops an orange band that yields 0.92 g (49%) of product after evaporation of the eluent.

Anal. Calcd. for $C_{27}H_{26}O_3Ru_2$: C, 54.5; H, 3.4. Found: C, 54.0; H, 3.4.

Properties

The compound is a red, crystalline, air-stable solid, soluble in common organic solvents. The IR spectrum (in CH_2Cl_2) exhibits bands at 1978 (s), 1803 (s), and 1731 (m) cm^{-1} due to the carbonyl and ketone groups. The ^{1}H NMR spectrum (in C_5D_5N) shows singlets at $\delta = 5.28$ and 5.56 for the η^5-cyclopentadienyl ligands and a multiplet centered at $\delta = 7.30$ for the phenyl protons.

The compound reacts readily in boiling toluene with alkynes (HC≡CH, HC≡CMe, MeC≡CMe, HC≡CPh, MeC≡CPh, MeOOCC≡CCOOMe) to form the analogs $[Ru_2(CO)(\mu\text{-}CO)\{\mu\text{-}\eta^1:\eta^3\text{-}C(O)CR^1CR^2\}(\eta^5-C_5H_5)_2]$ in good yields.[8] Likewise, reaction of the compound in refluxing toluene with CH_2=CH_2, SO_2, or $P(OMe)_3$, leads to substitution products of the formula $[Ru_2(CO)_3(L)(\eta^5-C_5H_5)_2]$.[9] With allene an unusual reaction occurs to yield $[Ru(CO)(\eta^5-C_5H_5)\{\eta^3-C_3H_4\text{-}2\text{-}Ru(CO)_2(\eta^5-C_5H_5)\}]$.[13] Ylides provide a general route to μ-alkylidene compounds, exemplified below.

C. μ-CARBONYL-μ-METHYLENE-BIS[CARBONYL-(η^5-CYCLOPENTADIENYL)RUTHENIUM]

$[Ru_2(CO)(\mu\text{-}CO)\{\mu\text{-}\eta^1:\eta^3\text{-}C(O)C_2Ph_2\}(\eta^5-C_5H_5)_2](I) + Ph_3PCH_2 \longrightarrow$

$[Ru_2(CO)_2(\mu\text{-}CO)(\mu\text{-}CH_2)(\eta^5-C_5H_5)_2] + PhC_2Ph + PPh_3$

Procedure

A two-necked, 250-mL flask equipped with a nitrogen by-pass, a reflux condenser, and a magnetic stirring bar is charged with $[Ru_2(CO)(\mu\text{-}CO)\{\mu\text{-}\eta^1:\eta^3\text{-}C(O)C_2Ph_2\}(\eta^5-C_5H_5)_2]$ (0.50 g, 0.84 mmol) and 100 mL of dry, deoxygenated toluene. A solution of Ph_3PCH_2 (3 mmol), prepared by the literature method,[14] in 20 mL of toluene is added to the flask and the mixture is heated at reflux for 0.5 hr, during which time a color change from orange-red to yellow is observed. The solvent is removed under reduced pressure and the residue is chromatographed on an alumina (Brockman grade 2) column (20 × 3 cm). Elution with dichloromethane–hexane (2:3) develops a yellow band that yields 0.25 g (70%) of yellow crystalline product, and an orange band that yields 0.055 g (10%) of the side product $[Ru_2(CO)(PPh_3)(\mu\text{-}CO)_2(\eta^5\text{-}C_5H_5)_2]$ as orange crystals.[10]

Anal. Calcd. for $C_{14}H_{12}O_3Ru_2$: C, 39.1; H, 2.8. Found: C, 39.1; H, 2.9. Yields of 30 and 35% were obtained by the checkers.

Properties

The compound is an air-stable, yellow crystalline solid and is readily soluble in common organic solvents. It is isolated as a mixture of cis and trans isomers that interconvert in solution too rapidly to allow their separation by chromatography. The IR spectrum (in CH_2Cl_2) of the mixture of isomers show bands at 1985 (s), 1941 (m), and 1781 (m) cm^{-1} due to the carbonyl ligands. The ^{1}H NMR spectrum (in CDCl$_3$) exhibits singlet resonances for the trans isomer at $\delta = 5.32$ (cyclopentadienyl protons) and 8.44 (μ-methylene protons) and for the cis isomer at $\delta = 5.24$ (C_5H_5) and at 7.52 and 9.16 (μ-CH$_2$).

Related μ-alkylidene derivatives, $[Ru_2(CO)_2(\mu\text{-}CO)(\mu\text{-}CHR)(\eta^5\text{-}C_5H_5)_2]$ (R = Me, Et, Ph, CH=CH$_2$) can be prepared in good to moderate yields by the analogous reaction of $[Ru_2(CO)(\mu\text{-}CO)\{\mu\text{-}\eta^1\text{:}\eta^3\text{-}C(O)C_2Ph_2\}(\eta^5\text{-}C_5H_5)_2]$ with Ph$_3$PCHR.[10] In a similar manner, the reactions of thermally robust diazoalkanes [CH(COOEt)N$_2$ or Ph$_2$CN$_2$] with $[Ru_2(CO)(\mu\text{-}CO)\{\mu\text{-}\eta^1\text{:}\eta^3\text{-}C(O)C_2Ph_2\}(\eta^5\text{-}C_5H_5)_2]$ in refluxing toluene produce the corresponding μ-alkylidene derivatives.[10]

The μ-methylene complex $[Ru_2(CO)_2(\mu\text{-}CO)(\mu\text{-}CH_2)(\eta^5\text{-}C_5H_5)_2]$ may be prepared directly from $[Ru_2(CO)_4(\eta^5\text{-}C_5H_5)_2]$ and Li[BHEt$_3$] (see Section F).

D. μ-CARBONYL-μ-ETHENYLIDENE-BIS[CARBONYL-(η^5-CYCLOPENTADIENYL)RUTHENIUM]

$[Ru_2(CO)_4(\eta^5\text{-}C_5H_5)_2]$ + MeLi $\longrightarrow$
$$[Ru_2(CO)_3\{C(O)Me\}(\eta^5\text{-}C_5H_5)_2]^- + Li^+$$

$[Ru_2(CO)_3\{C(O)Me\}(\eta^5\text{-}C_5H_5)_2]^- + 2H^+ \longrightarrow$
$$[Ru_2(CO)_2(\mu\text{-}CO)(\mu\text{-}CMe)(\eta^5\text{-}C_5H_5)_2]^+ + H_2O$$

$[Ru_2(CO)_2(\mu\text{-}CO)(\mu\text{-}CMe)(\eta^5\text{-}C_5H_5)_2]^+ + NEt_3 \longrightarrow$
$$[Ru_2(CO)_2(\mu\text{-}CO)(\mu\text{-}CCH_2)(\eta^5\text{-}C_5H_5)_2] + NHEt_3^+$$

■ **Caution.** *Methyllithium diethyl ether solutions are pyrophoric in air and must be handled under an inert atmosphere. It is advisable to use a hood.*

Procedure

A Schlenk flask equipped with a stopper, a nitrogen by-pass, and a magnetic stirring bar is charged with tetracarbonylbis(η^5-cyclopentadienyl)diruthenium (0.79 g, 1.78 mmol) and 100 mL of dry, deoxygenated THF. Halide-free methyllithium in diethyl ether (2.0 mL of 1 M solution) is added by syringe and the solution is stirred at room temperature for 45 min. The reaction mixture is cooled to $-78°$ and an excess of H[BF$_4$]·OEt$_2$ (~2 mL) is added, giving [Ru$_2$(CO)$_2$(μ-CO)(μ-CMe)(η^5-C$_5$H$_5$)$_2$][BF$_4$] as an orange precipitate. The mixture is allowed to warm to room temperature over ~1 hr, an excess of NEt$_3$ (5 mL) added, and then the whole is stirred for a further 10 min. The solvent is removed under pressure and the oily residue extracted with dichloromethane (~200 mL). The extract is washed with water (~200 mL) to remove ammonium salt, the water layer is back-extracted with dichloromethane (~200 mL), and the solvent is evaporated from the combined dichloromethane solutions. The residue is chromatographed on an alumina column. Elution with dichloromethane–hexane (2:3) develops two yellow bands that yield, in turn, trans and cis isomers of the product in a combined yield of 0.60 g (76%).

Anal. Calcd. for C$_{15}$H$_{12}$O$_3$Ru$_2$: C, 40.7; H, 2.7. Found: C, 40.7; H, 2.8.

Properties

The compound is a yellow, crystalline, air-stable solid, soluble in common organic solvents. The IR spectrum (in CH$_2$Cl$_2$) of the trans isomer shows carbonyl bands at 1953 (s) and 1793 (m) cm^{-1}; the cis isomer shows carbonyl bands at 1994 (s), 1951 (w), and 1788 (m) cm^{-1}. The ^{1}H NMR spectrum (in CDCl$_3$) of the trans product exhibits singlets at $\delta = 5.26$ for the η^5-C$_5$H$_5$ protons and at 6.37 for the vinylidene protons; the cis product shows singlets at $\delta = 5.20$ (η^5-C$_5$H$_5$) and 6.27 (μ-CCH$_2$).

E. μ-CARBONYL-μ-ETHYLIDYNE-BIS[CARBONYL-(η^5-CYCLOPENTADIENYL)RUTHENIUM] TETRAFLUOROBORATE

$$[Ru_2(CO)_2(\mu\text{-}CO)(\mu\text{-}CCH_2)(\eta^5\text{-}C_5H_5)_2] + H[BF_4] \longrightarrow$$
$$[Ru_2(CO)_2(\mu\text{-}CO)(\mu\text{-}CMe)(\eta^5\text{-}C_5H_5)_2][BF_4]$$

Procedure

The reaction is carried out in a 250-mL, round-bottomed flask equipped with a magnetic stirring bar. A nitrogen atmosphere was used, but is not strictly necessary. A sample of μ-carbonyl-μ-ethenylidene-bis[carbonyl(η⁵-cyclopentadienyl)ruthenium] (0.20 g, 0.45 mmol) is dissolved in 100 mL of dichloromethane. A few drops of $H[BF_4] \cdot OEt_2$ are added to the stirred solution, causing an immediate color change from yellow to orange. The solvent is removed under reduced pressure and the residue washed three times with 30-mL portions of diethyl ether. Crystallization from dichloromethane affords 0.22 g (92%) of product.

Anal. Calcd. for $C_{15}H_{13}BF_4O_3Ru_2$: C, 34.0; H, 2.5. Found: C, 34.2; H, 2.7.

Properties

The compound is an orange crystalline solid that is stable under vacuum, but slowly decomposes in the air. It is soluble in dichloromethane and acetone, and solutions may be handled briefly in air. The IR spectrum (in CH_2Cl_2) shows bands at 2049 (s), 2014 (w), and 1859 (m) cm^{-1} due to the carbonyl ligands. The ^1H NMR spectrum (in CD_2Cl_2) shows singlets at $\delta = 4.62$ for the ethylidyne protons and at 5.72 for the η⁵-cyclopentadienyl protons.

F. μ-CARBONYL-μ-ETHYLIDENE-BIS[CARBONYL-(η⁵-CYCLOPENTADIENYL)RUTHENIUM]

$[Ru_2(CO)_4(\eta^5-C_5H_5)_2]$ + MeLi $\longrightarrow$

$$[Ru_2(CO)_3\{C(O)Me\}(\eta^5-C_5H_5)_2]^- + Li^+$$

$[Ru_2(CO)_3\{C(O)Me\}(\eta^5-C_5H_5)_2]^- + 2H^+ \longrightarrow$

$$[Ru_2(CO)_2(\mu-CO)(\mu-CMe)(\eta^5-C_5H_5)_2]^+ + H_2O$$

$[Ru_2(CO)_2(\mu-CO)(\mu-CMe)(\eta^5-C_5H_5)_2]^+ + H^- \longrightarrow$

$$[Ru_2(CO)_2(\mu-CO)(\mu-CHMe)(\eta^5-C_5H_5)_2]$$

■ **Caution.** *Methyllithium diethyl ether solutions are pyrophoric in air and must be handled under an inert atmosphere. It is advisable to use a hood.*

Procedure

A Schlenk flask equipped with a stopper, a nitrogen by-pass, and a magnetic stirring bar is charged with tetracarbonylbis(η^5-cyclopentadienyl)diruthenium (1.00 g, 2.25 mmol) and 40 mL of dry, deoxygenated THF. Halide-free methyllithium in diethyl ether (2.3 mL of 1 M solution) is added by syringe and the mixture is stirred for 1 hr. The reaction is cooled to $-78°$, an excess of H[BF$_4$]·OEt$_2$ (~2 mL) added, and the mixture stirred at this temperature for 30 min. An excess of Na[BH$_4$] (0.5 g, 13 mmol) is added and the mixture is allowed to warm to room temperature over 30 min. The solvent is evaporated at reduced pressure, the residue extracted with dichloromethane (~100 mL), and the extracts filtered through a short alumina column. The filtrate is chromatographed on another alumina (Brockman grade 2) column (20 × 3 cm). Elution with dichloromethane–hexane (1:1) gives a single yellow band which yields 0.89 g (89%) of product.

Anal. Calcd. for C$_{15}$H$_{14}$O$_3$Ru$_2$: C, 40.5; H, 3.2. Found: C, 40.4; H, 3.2.

Properties

The compound is a yellow, air-stable crystalline solid, soluble in common organic solvents. The IR spectrum (in CH$_2$Cl$_2$) of the mixture of cis and trans isomers obtained shows bands at 1974 (s), 1933 (m), and 1776 (m) cm^{-1} due to the carbonyl ligands. The ^{1}H NMR spectrum (in CDCl$_3$) shows resonances for the trans isomer at δ = 3.16 (d, J = 7 Hz, μ-CHMe), 5.18 (s, C$_5$H$_5$), 5.24 (s, C$_5$H$_5$), and 10.14 (q, J = 7 Hz, μ-CHMe), and for the cis isomer at δ = 3.04 (d, J = 7 Hz, μ-CHMe), 5.18 (s, 2C$_5$H$_5$), and 10.94 (q, J = 7 Hz, μ-CHMe).

A related sequence, involving the action of Li[BHEt$_3$] on [Ru$_2$(CO)$_4$-(η^5-C$_5$H$_5$)$_2$], produces the previously described methylene complex [Ru$_2$(CO)$_2$(μ-CO)(μ-CH$_2$)(η^5-C$_5$H$_5$)$_2$] in excellent yield.[15]

References

1. E. L. Muetterties and J. Stein, *Chem. Rev.*, **79**, 479 (1979).
2. See, for example, W. A. Herrmann, *Angew. Chem., Int. Ed. (Engl.)*, **21**, 117 (1982).
3. E. O. Fischer and A. Vogler, *Z. Naturforsch.*, **B17**, 421 (1962).
4. T. Blackmore, M. I. Bruce, and F. G. A. Stone, *J. Chem. Soc. (A)*, **1968**, 2158.
5. A. P. Humphries and S. A. R. Knox, *J. Chem.Soc., Dalton Trans.*, **1975**, 1710.
6. A. F. Dyke, S. R. Finnimore, S. A. R. Knox, P. J. Naish, A. G. Orpen, G. H. Riding, and G. E. Taylor, *Reactivity of Metal–Metal Bonds*, M. H. Chisholm (ed.), (ACS Symposium Series, Vol. 155) Washington, Vol. 1981, p. 259.
7. S. A. R. Knox, *Pure Appl.Chem.*, **56**, 81 (1984) and references therein.

8. A. F. Dyke, S. A. R. Knox, P. J. Naish, and G. E. Taylor, *J. Chem. Soc., Dalton Trans.,* **1982,** 1297.
9. D. L. Davies, A.F. Dyke, S. A. R. Knox, and M. J. Morris, *J. Organomet. Chem.,* **215,** C30 (1981).
10. D. L. Davies, S. A. R. Knox, K. A. Mead, M. J. Morris, and P. Woodward, *J. Chem. Soc., Dalton Trans.,* **1984,** 2293.
11. R. E. Colborn, D. L. Davies, A. F. Dyke, A. Endesfelder, S. A. R. Knox,A. G. Orpen, and D. Plaas, *J. Chem. Soc., Dalton Trans.,* **1983,** 2661.
12. M. I. Bruce and C. H. Hameister, see *Comprehensive Organometallic Chemistry,* Vol. 4, G. Wilkinson, F. G. A. Stone, and E. W. Abel, (eds.), Pergamon Press, 1982, p. 664.
13. R. E. Colborn, A. F. Dyke, S. A. R. Knox, K. A. Mead, and P. Woodward, *J. Chem. Soc., Dalton Trans.,* **1983,** 2099.
14. H. Schmidbaur, H. Stühler, and W. Vornberger, *Chem. Ber.,* **105,** 1084 (1972).
15. D. H. Berry and J. E. Bercaw, private communication.

45. DI-μ-IODO-BIS(TRICARBONYLOSMIUM), BIS(TETRACARBONYLIODOOSMIUM), AND DICARBONYLIODO(η^5-CYCLOPENTADIENYL)OSMIUM

Submitted by STEVEN ROSENBERG,* ALBERT W. HERLINGER,†
WAYNE S. MAHONEY,* and GREGORY L. GEOFFROY*
Checked by ROBERT T. HEMBRE,† K. R. BIRDWHISTELL,†
and J. NORTON†

Metal carbonyl–halide complexes are valuable synthetic reagents for the preparation of many interesting mononuclear and polynuclear compounds. Osmium reagents include such complexes as $Os(CO)_4X_2$,[1] $Os_2(CO)_8X_2$,[2] and $Os_2(CO)_6(\mu\text{-}X)_2$,[3] where X = Cl, Br, and I. The binuclear compounds serve as useful synthetic materials for the preparation of bi- and polynuclear organometallic compounds. The utility of the two binuclear reagents, however, has been limited by the lack of convenient routes for their synthesis. Previously the title compounds have been obtained in low yields by reaction of CF_3I or I_2 with $Os_3(CO)_{12}$ in benzene in sealed tubes over several days.[4] The $Os_2(CO)_8I_2$ complex has also been reported to be obtainable in good yield by reacting $Os_3(CO)_{12}$ with iodine under controlled conditions,[3] but the specific details of this synthesis were not given. The same authors report that refluxing $Os_2(CO)_8I_2$ in heptane yields $Os_2(CO)_6(\mu\text{-}I)_2$ in high yield.[3] Reported herein are details of the high-yield syntheses of $Os_2(CO)_6(\mu\text{-}I)_2$

*Department of Chemistry, The Pennsylvania State University, University Park, PA 16802.
†Department of Chemistry, Loyola University of Chicago, Chicago, IL 60626.
‡Department of Chemistry, Colorado State University, Fort Collins, CO 80523.

and $Os_2(CO)_8I_2$ by careful control of reaction variables in the $Os_3(CO)_{12}$ + I_2 reaction.

The thermal reaction of C_5R_5H (R = H, CH_3) with $Os_2(CO)_6(\mu\text{-}I)_2$ represents a convenient synthesis of the cyclopentadienyl complexes $Os(\eta^5\text{-}C_5R_5)(CO)_2I$, and details of these preparations are given. The complex $OsCp(CO)_2I$ (Cp = C_5H_5) has been previously prepared in 51% yield by the thermal reaction of TlC_5H_5 with $Os(CO)_4Br_2$ to give a mixture of $OsCp(CO)_2H$ and $Os_2Cp_2(CO)_4$, which was subsequently treated with I_2 to yield the desired product.[5] The complex $OsCp^*(CO)_2I$ (Cp* = C_5Me_5) has been prepared in 38% yield by the thermal reaction of $Os_3(CO)_{12}$ with C_5Me_5H to give $OsCp(CO)_2H$, which was then treated with I_2.[5]

A. HEXACARBONYL-DI-μ-IODODIOSMIUM(I)

$$2Os_3(CO)_{12} + 3I_2 \longrightarrow 3Os_2(CO)_6(\mu\text{-}I)_2 + 6CO \qquad (1)$$

■ **Caution.** *The reaction should be performed in a well-ventilated fume hood since toxic carbon monoxide is liberated. The Carius tube should be wrapped with a wire screen and placed behind a safety shield.*

Procedure

Dodecacarbonyltriosmium, $Os_3(CO)_{12}$, used as starting material for this preparation may be synthesized as described in Vol 21 of *Inorganic Syntheses*.[6]

A 300 mL resealable Carius tube fitted with a Teflon vacuum stopcock is charged with $Os_3(CO)_{12}$ (0.45 g, 0.5 mmol), I_2 (0.19 g, 0.75 mmol), and 25 mL of dry toluene. The Carius tube is evacuated to ~0.01 torr and heated in an oil bath at 175–180° for 24 hr. The tube should be immersed as fully as possible in the oil bath. After cooling to room temperature, the Carius tube is opened in air and the yellow solution transferred to a 250-mL round-bottom flask. Anhydrous methanol (75 mL) is then added to facilitate removal of the solvent by rotary evaporation as a 3:1 azeotrope. The product obtained at this point may be pure enough for many synthetic purposes and the yield (0.57 g, 0.71 mmol) is nearly quantitative (95%). Further purification can be achieved by chromatography on Florisil using a 4 in. × ¾ in. column with hexane as the eluant. The only material that elutes is yellow $Os_2(CO)_6(\mu\text{-}I)_2$. Evaporation of solvent from this band followed by drying in a stream of nitrogen gives the compound as a yellow microcrystalline solid (0.48 g, 80%, mp 87–89°). If desired, the compound can be further purified by sublimation (0.01 torr, 60°).

The yield of $Os_2(CO)_6(\mu\text{-}I)_2$ is highly dependent on reaction conditions, particularly temperature. Consequently, care must be exercised to minimize temperature gradients within the reaction vessel. The Carius tube should be immersed as fully as possible in the oil bath, exposed portions of the tube covered with foil, and the temperature of the bath maintained in the 175°–180° range. If the oil bath is allowed to drop below this temperature range, for example, to 165°–175°, mixtures of $Os_2(CO)_6(\mu\text{-}I)_2$ (~70%) and $Os_2(CO)_8I_2$ (~30%) are obtained. This mixture can be upgraded to give essentially quantitative formation of $Os_2(CO)_6(\mu\text{-}I)_2$ by reheating in an evacuated Carius tube at 175–180° for 12 hr. Alternatively, the mixture can be conveniently separated by chromatography on Florisil using a 4 in. × ¾ in. column. Elution with hexane gives $Os_2(CO)_6(\mu\text{-}I)_2$ and subsequent elution with 1:1 (v/v) hexane–dichloromethane yields $Os_2(CO)_8I_2$ as a yellow solid.

$$Os_2(CO)_8I_2 \xrightarrow[\text{reflux}]{\Delta} Os_2(CO)_6(\mu\text{-}I)_2 + 2CO \qquad (2)$$

■ **Caution.** *The reaction should be performed in a well-ventilated fume hood since toxic carbon monoxide is liberated. The Carius tube should be wrapped with a wire screen and placed behind a safety shield.*

Procedure

This procedure is similar to that described by Sutton and co-workers.[3] A sample of $Os_2(CO)_8I_2$ (0.10 g, 0.12 mmol), prepared as described in the following synthesis, is suspended in 25 mL of dry nitrogen saturated heptane in a 100-mL, three-neck, round-bottom flask fitted with a gas inlet, a reflux condenser, and a magnetic stirring bar. The reaction mixture is stirred while heating to reflux under dry oxygen-free nitrogen. The suspended solid gradually dissolves as the heptane is heated to boiling and the resultant yellow solution is refluxed under nitrogen for 45 min. After the solution has cooled to room temperature, the solvent is removed under reduced pressure on a rotary evaporator. The resultant yellow solid is dissolved in hexane and purified by chromatography on Florisil as described in the previous preparation. The complex $Os_2(CO)_6(\mu\text{-}I)_2$ is obtained as a yellow-orange microcrystalline solid (0.077 g, 80%).

Properties

The orange-yellow crystalline product is air stable. It readily dissolves in both polar and nonpolar solvents giving yellow moisture sensitive, but thermally stable, solutions. Full spectroscopic and structural data have been reported,[3] and the compound may be conveniently characterized by its IR spectrum in the carbonyl stretching region: ν_{CO}(hexane, abs. mode) = 2098 (m), 2069 (s), 2016 (s), 2010 (s), and 2004 (sh). The mass spectrum shows a parent ion at m/e 806 (^{192}Os) and fragment ions corresponding to progressive loss of 6 CO groups. The pure compound melts in the 87–89° range.

B.   BIS[TETRACARBONYLIODOOSMIUM(I)]

$$Os_2(CO)_6(\mu\text{-}I)_2 + 2CO \longrightarrow Os_2(CO)_8I_2 \tag{3}$$

■ **Caution.** *This synthesis should be performed in a well-ventilated fume hood since toxic carbon monoxide is involved. The glass pressure bottle should be wrapped in a wire screen and placed behind a safety shield.*

Procedure

A 100-mL glass pressure bottle, Fig. 1, containing a magnetic stirring bar is charged with $Os_2(CO)_6(\mu\text{-}I)_2$ (1.25 g, 1.56 mmol) and hexane (50 mL).* The reaction vessel is then pressurized to 90 psi with carbon monoxide, vented, and repressurized to 90 psi with CO. Stirring is continued at room temperature for 36 hr. During this time, $Os_2(CO)_8I_2$ gradually precipitates from solution as a yellow microcrystalline solid. The reaction vessel is vented in a fume hood in air. The crystals are collected by filtration, washed with several small portions of hexane, and air dried (0.74 g, 55%, mp 133–136°). Further material can be obtained by chromatography of the yellow filtrate on Florisil (4 in. × $\frac{3}{4}$ in. column). After loading the filtrate on the column, elution with hexane gives a yellow band of unreacted $Os_2(CO)_6(\mu\text{-}I)_2$. Further elution with 1:1 (v/v) hexane–dichloromethane gives a yellow band of $Os_2(CO)_8I_2$. The solvent is removed from the latter on a rotary evaporator. The resultant solid is transferred to a fritted glass filter and washed with 10 mL of hexane to give an additional 0.40 g of $Os_2(CO)_8I_2$ (total yield 85%).

Properties

The yellow air-stable crystalline product readily dissolves in polar solvents, but is considerably less soluble in nonpolar solvents than $Os_2(CO)_6(\mu\text{-}I)_2$. Solutions of $Os_2(CO)_8I_2$ are thermally stable below ~60%, but above this

---

*Alternatively, an autoclave can be used for this procedure employing 700 psi of CO pressure at 23° for 18 hr.

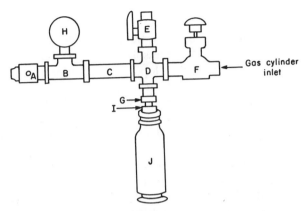

Fig. 1 Sources and description of apparatus. (I) Pittsburgh Valve and Fitting Company: A, Nupro safety relief valve (100 psi release) 3-6R-4M-100; B, Cajon $\frac{1}{4}$ in. Tee B-4-T; C, Cajon $\frac{1}{4}$ in. Hex long nipple B-4-HLN(2-$\frac{1}{4}$ in.); D, Cajon $\frac{1}{4}$ in. Cross B-4-CS; E, Whitey Ball Valve B-43M-S4; F, Whitey $\frac{1}{4}$ in. Forged Body Valve B-1VM4-S4; G, Cajon hex reducing nipple, $\frac{1}{4}$ to $\frac{1}{8}$ in. B-4-HRN-2. (II) Mattheson Gas Products: H, Standard Guage (2–3% accuracy) 0-100 psig 63-3112. (III) Lab Crest Scientific: I, Needle Valve adapter 110-957; Rubber washer 110-973; J, Aerosol reaction vessel, 3 oz 110-023-0003.

temperature the compound loses CO to yield $Os_2(CO)_6(\mu\text{-}I)_2$. Spectroscopic data have been reported for this compound,[4] and it may be conveniently characterized by its IR spectrum in the carbonyl stretching region: ν_{CO} (hexane, abs mode) = 2112 (s), 2076 (s), 2063 (s), 2060 (vs), 2049 (s), and 2029 (s) cm^{-1}. The mass spectrum shows the parent molecular ion at m/e 862 (^{192}Os) and fragment ions corresponding to progressive loss of 8 CO groups. The pure material melts at 133–136°.

C. DICARBONYL(η^5-CYCLOPENTADIENYL)IODOOSMIUM AND DICARBONYLIODO(η^5PENTAMETHYL-CYCLOPENTADIENYL)OSMIUM

$$Os_2(CO)_6(\mu\text{-}I)_2 + (C_5R_5H) \xrightarrow{\Delta}$$
$$2Os(\eta^5\text{-}C_5R_5)(CO)_2I + 2\,CO + H_2 \quad (4)$$
$$R = H,\ Me$$

■ **Caution.** *The reaction should be peformed in a well-ventilated fume hood since toxic carbon monoxide is liberated. The Carius tube should be wrapped with a wire screen and placed behind a safety shield.*

Procedure

$Os_2(CO)_6(\mu\text{-}I)_2$ is generated *in situ* in the manner described in Section A but substituting benzene for toluene solvent. The Carius tube is opened in air and charged with $\frac{1}{2}$ mL of dicyclopentadiene (Aldrich Chemical Co.). After evacuation to 0.01 torr, the Carius tube is sealed and heated in an oil bath at 175–180° for 72 hr. It is important that the tube be immersed as fully as possible in the oil bath and the temperature maintained above 175°. After cooling and opening in air, solvent is removed from the yellow solution by rotary evaporation to yield a yellow solid. This is dissolved in 25 mL of CH_2Cl_2 followed by addition of 50 mL of hexane. The solution is then concentrated to 25 mL and placed in a freezer at $-10°$ overnight. The yellow solid that precipitates is collected on a medium porosity frit, washed with 10 mL of hexane, and dried *in vacuo* to afford 899 mg (2.05 mmol) of $Cp(CO)_2OsI$, in 62% yield. The filtrates from above may be concentrated by rotary evaporation and the recrystallization procedure repeated to afford another 30% (435 mg, 0.99 mmol) of $OsCp(CO)_2I$. The overall yield varies from 86 to 92%. The pentamethylcyclopentadienyl analog can be prepared similarly in 85–88% yield by substituting 0.6 mL of C_5Me_5H (Alfa Products Company or ref. 7) for $(C_5H_6)_2$.

Properties

Both $OsCp(CO)_2I$ and $OsCp^*(CO)_2I$ are air stable yellow solids. They readily dissolve in polar solvents and are very slightly soluble in hexane. Full spectroscopic data have been reported for both complexes,[5] and they may be conveniently characterized by their IR spectra in the carbonyl stretching region: ν_{CO}(hexane, abs modes), $OsCp(CO)_2I = 2039$ (s), 1988 (s) cm^{-1}; $Os(Cp^*)(CO)_2I = 2020$ (s), 1968 (s) cm^{-1}.

References

1. F. L'Eplattenier and F. Calderazzo, *Inorg. Chem.*, **6**, 2092 (1967).
2. K. M. Motyl, J. R. Norton, C. K. Schauer, and O. P. Anderson, *J. Am. Chem. Soc.*, **104**, 7325 (1982).
3. E. E. Sutton, M. L. Niven, and J. R. Moss, *Inorg. Chem. Acta*, **70**, 207 (1983).
4. M. I. Bruce, M. Cooke, M. Green, and D. J. Westlake, *J. Chem. Soc. (A)*, 1969, 987.
5. J. K. Hoyano, C. J. May, and W. A. G. Graham, *Inorg. Chem.*, **21**, 3095 (1982).
6. B. F. G. Johnson and J. Lewis, *Inorg. Synth.*, **13**, 92 (1972).
7. (a) F. X. Kohl and P. Jutzi, *J. Organomet. Chem.*, **243**, 119 (1983). (b) J. M. Manriquez, P. J. Fagan, L. D. Schertz, and T. J. Marks, *Inorg. Synth.*, **21**, 181 (1982).

46.   μ-NITRIDO-BIS(TRIPHENYLPHOSPHORUS)(1+)-μ-CARBONYL-DECACARBONYL-μ-HYDRIDOTRIOSMATE(1−)

Submitted by K. BURGESS* and R. P. WHITE*
Checked by S. BASSNER,† G. L. GEOFFROY,† R. L. GRAY,‡ and D. J. DARENSBOURG‡

The starting material for most of the osmium cluster chemistry published to date, $Os_3(CO)_{12}$,[1] is quite stable and relatively unreactive. Hydroxide ions remove one carbonyl ligand from $Os_3(CO)_{12}$ and a triosmium anion results that is far more reactive than the cluster from which it is produced. Indeed, this anion, $[Os_3(\mu\text{-}H)(\mu\text{-}CO)(CO)_{10}]^-$, combines under mild conditions with a range of inorganic,[2,3] organic,[4,5] and organometallic,[6,7] electrophiles providing a route to functionalized triosmium complexes and mixed-metal clusters. A reliable and convenient modification of the original synthesis of $[Os_3(\mu\text{-}H)(\mu\text{-}CO)(CO)_{10}]^-$ (ref 8) is presented here.

$$Os_3(CO)_{12} + KOH + (PPh_3)_2NCl \longrightarrow$$
$$[(PPh_3)_2N][Os_3(\mu\text{-}CO)(CO)_{10}] + KCl + CO_2$$

■ **Caution.** *Owing to the high toxicity of carbon monoxide and osmium carbonyls these reactions should be carried out in an efficient fume hood.*

Powdered samples of 0.454 g (0.5 mmol) of $Os_3(CO)_{12}$ (ref. 1) and 1.403 g (25 mmol) of KOH are placed in a 50-mL Schlenk tube with a magnetic stirrer. The reaction vessel is sealed with a septum and evacuated–filled with nitrogen three times to remove oxygen. The Schlenk tube and its contents are cooled in liquid nitrogen, 25 mL of methanol (distilled from magnesium and taken directly from a still collector vessel under dinitrogen) is added by syringe, and the mixture is degassed using three freeze–thaw cycles. The reaction mixture is then warmed to 25° and vigorously stirred for 6 hr during which time the solution changes from yellow to red. A degassed solution of 0.32 g (0.56 mmol) of $(PPh_3)_2NCl$ (Aldrich Chemical

*University Chemical Laboratory, Cambridge, CB2 1EW, United Kingdom. Address correspondence to Kevin Burgess, Chemistry, Box 1892, Rice University, Houston, TX 77251.
†Department of Chemistry, The Pennsylvania State University, University Park, PA 16802.
‡Department of Chemistry, Texas A&M University, College Station, TX 77843.

Co.) in 3 mL of methanol is added all at once and the stirring is continued for 15 min. Degassed distilled water (~2 mL required) is carefully added dropwise, with stirring, over 5 min until a red precipitate appears and persists on stirring for 2 min. The water should be added with çare since too much will precipitate some $(PPh_3)_2NCl$. The suspension is stirred further for 1 hr. The precipitate is then filtered in air, washed with 5 mL of distilled water and 5 mL of cold methanol, and dried *in vacuo* overnight to give 58%, 0.414 g, (0.292 mmol) of μ-nitrido-bis(triphenylphosphorus)(1 +)μ-carbonyl-decarbonyl-μ-hydrido-triosmate(1 −).

Anal. Calcd. for $C_{47}H_{31}NO_{11}P_2Os_3$: C, 39.80; H, 2.20; N, 0.99; P, 4.37. Found: C, 40.05, H, 2.28; N, 1.10; P, 4.45.

Properties

μ-Nitrido-bis(triphenylphosphorus)(1 +) μ-carbonyl-decarbonyl-μ-hydrido-triosmate(1 −), $[(PPh_3)_2N][Os_3(\mu\text{-}H)(\mu\text{-}CO)(CO)_{10}]$ is obtained as a light red powder. The solid is air stable but solutions decompose in ~5 min when exposed to the atmosphere. It is soluble in THF, dichloromethane, acetonitrile, methanol, and diethyl ether, and insoluble in hydrocarbon solvents. The IR spectrum of the compound contains four CO stretching vibrations for terminal carbonyls and one bridging carbonyl stretch (CH_2Cl_2, cm^{-1}): 2038 (w), 2021 (s), 1996 (s), 1951 (ms), and 1667 (w), respectively. The ^1H NMR spectrum (80 MHz, chloroform-d_1, δ in ppm downfield from TMS, ambient) shows a broad multiplet at 7.5, due to the protons attached to the aromatic rings in the cation, and a sharp metal hydride signal at − 13.8.

References

1. B. F. G. Johnson and J. Lewis, *Inorg. Synth.*, **13**, 93 (1972).
2. B. F. G. Johnson, P. R. Raithby, and C. Zuccaro, *J. Chem. Soc. Dalton Trans.*, **1980**, 99.
3. B. F. G. Johnson, J. Lewis, P. R. Raithby, and S. W. Sankey, *J. Organomet. Chem.*, **228**, 135 (1982).
4. J. B. Keister, *J. Chem. Soc. Chem. Commun.*, **1979**, 214.
5. C. E. Kampe, N. M. Boag, and H. D. Kaesz, *J. Am. Chem. Soc.*, **105**, 2896 (1983).
6. B. F. G. Johnson, D. A. Kaner, J. Lewis, and P. R. Raithby, *J. Organomet. Chem.*, **215**, C33 (1981).
7. M. Fajardo, H. D. Holden, B. F. G. Johnson, J. Lewis, and P. R. Raithby, *J. Chem. Soc. Chem. Commun.*, **1984**, 24.
8. C. R. Eady, B. F. G. Johnson, J. Lewis, and M. C. Malatesta, *J. Chem. Soc. Dalton Trans.*, **1978**, 1358.

47. DECACARBONYL-(η⁵-CYCLOPENTADIENYL)DIHYDRIDOCOBALTTRIOSMIUM, CoOs₃(μ-H)₂(η⁵-C₅H₅)(μ-CO)(CO)₉; NONACARBONYL-(η⁵-CYCLOPENTADIENYL)TRIHYDRIDO- AND (η⁵-CYCLOPENTADIENYL)TETRAHYDRIDOCOBALTTRIOSMIUM, CoOs₃(μ-H)₃(η⁵-C₅H₅)(CO)₉ AND CoOs₃(μ-H)₄(η⁵-C₅H₅)(CO)₉

Submitted by DENG-YANG JAN,* TIM J. COFFY,* WEN-LIANG HSU,*
and SHELDON G. SHORE*
Checked by B. HANDWERKER† and GREGORY L. GEOFFROY†

A general route to triosmium based mixed metal, tetranuclear clusters involves reactions of the cluster $Os_3(\mu\text{-H})_2(CO)_{10}$ with metal carbonyls under reaction conditions that cause formation of unsaturated metal carbonyl fragments generated under thermal[1-6] or photochemical[7] conditions. While $Os_3(\mu\text{-H})_2(CO)_{10}$ has usually been employed as the starting material it has also, apparently, been generated *in situ* from $Os_3(CO)_{12}$ in the formation of the mixed metal cluster in the presence of added H_2.[8,9] The effect of the presence or absence of added H_2 to the reaction system can determine the cluster products formed[4,5,8,10] and in the reactions of $[Ni(\eta^5\text{-}C_5H_5)CO]_2$ (ref. 8) and $[Mo(\eta^5\text{-}C_5H_5)(CO)_n]_2$ ($n = 2,3$)[10] the added H_2 is believed to assist in opening the metal–metal bond through formation of an intermediate, reactive hydride.

Here we describe the synthesis of $CoOs_3(\mu\text{-H})_2(\eta^5\text{-}C_5H_5)(\mu\text{-CO})(CO)_9$ from the reaction of $Os_3(\mu\text{-H})_2(CO)_{10}$ with $Co(\eta^5\text{-}C_5H_5)(CO)_2$, while two other clusters richer in hydrogen, $CoOs_3(\mu\text{-H})_3(\eta^5\text{-}C_5H_5)(CO)_9$ and $CoOs_3(\mu\text{-H})_4(\eta^5\text{-}C_5H_5)(CO)_9$ were synthesized in the presence of H_2.

Procedure

■ **Caution.** *Because of the toxicity of the CO evolved, all the reactions should be run in a well-ventilated hood.*

*Department of Chemistry, The Ohio State University, 140 West 18th Avenue, Columbus, OH 43210.
†Department of Chemistry, The Pennsylvania State University, University Park, PA 16802.

Materials

Toluene is dried, distilled, and stored in the presence of sodium benzophenone ketyl. The cluster $Os_3(\mu\text{-}H)_2(CO)_{10}$ is prepared according to a method in the literature[11] and $Co(\eta^5\text{-}C_5H_5)(CO)_2$ (Strem Chemicals) is distilled under high vacuum and stored at $-78°$.

A. DECACARBONYL(η^5-CYCLOPENTADIENYL)-DIHYDRIDOCOBALTTRIOSMIUM $CoOs_3(\mu\text{-}H)_2(\eta^5\text{-}C_5H_5)$-$(\mu\text{-}CO)(CO)_9$

$$Os_3(\mu\text{-}H)_2(CO)_{10} + Co(\eta^5\text{-}C_5H_5)(CO)_2 \longrightarrow$$
$$CoOs_3(\mu\text{-}H)_2(\eta^5\text{-}C_5H_5)(\mu\text{-}CO)(CO)_9 + 2CO \quad (1)$$

In a drybox containing a N_2 atmosphere a 10-mL quantity of dry toluene is syringed into a 50-mL round-bottom flask containing 51 mg $Os_3(\mu\text{-}H)_2(CO)_{10}$ (0.06 mmol) and a Teflon coated magnetic stirring bar. By means of a syringe 140 mg $Co(\eta^5\text{-}C_5H_5)(CO)_2$ (0.778 mmol) is then added to the solution. The flask is then capped with a solv-seal (grease-free) vacuum adapter and connected to a vacuum line. The flask is frozen at $-196°$ (liq N_2) and the N_2 gas is pumped away. After being allowed to warm up slowly to room temperature, the reaction mixture is stirred at $90°$ (under static vacuum). Consumption of $Os_3(\mu\text{-}H)_2(CO)_{10}$ is checked by TLC (SiO_2) and the CO produced in the reaction is pumped away from time to time ($\sim$ every 6–8 hr). The $Os_3(\mu\text{-}H)_2(CO)_{10}$ is consumed in 36 hr,[12] at which point the solution is cooled to room temperature and the solvent and excess $Co(\eta^5\text{-}C_5H_5)(CO)_2$ is pumped away under vacuum.[13] There remains in the reaction flask a dark green residue that is dissolved in a minimum amount of CH_2Cl_2 and subjected to TLC (SiO_2 preparative, 2 mm in thickness) using 1:4 toluene–hexane as the eluent. A green band (35 mg, 60%, $R_f = 0.48$) is collected and shown to be $CoOs_3(\mu\text{-}H)_2(\eta^5\text{-}C_5H_5)(\mu\text{-}CO)(CO)_9$.

Properties of $CoOs_3(\mu\text{-}H)_2(\eta^5\text{-}C_5H_5)(\mu\text{-}CO)(CO)_9$

This compound is a green solid. It is stable in the solid state, while slowly decomposing in solution when exposed to air over a long period of time. It is soluble in CH_2Cl_2, but only slightly soluble in hexane.

The IR spectrum (cyclohexane, ν_{CO}): 2095 (m), 2068 (vs), 2050 (vs), 2012 (vs), 2000 (sh), 1977 (m), 1968 (m), 1800 (s) cm^{-1}. Mass spectrum: parent ion is at m/e 981.8223. Consistent with $^1H_7$ $^{12}C_{15}$ $^{16}O_{10}$ ^{59}Co $^{192}Os_3$ (calcd. value is 981.8207). ^1H NMR spectra (90 MHz, CD_2Cl_2, $-80°$): $\delta = 5.47$ (s, 5H), -17.17 (s, 1H), -20.89 (s, 1H) ppm.

Anal. Calcd. 18.44C, 0.72 H. Found: 18.65 C, 0.77 H.

The structure of this compound has been determined by X-ray diffraction, and its structure in solution, inferred from ^{13}NMR spectrum at $-75°$, is consistent with that in the solid state.[6]

B. NONACARBONYL(η^5-CYCLOPENTADIENYL)TRIHYDRIDO AND TETRAHYDRIDO (η^5-CYCLOPENTADIENYL)-COBALTTRIOSMIUM, $CoOs_3(\mu\text{-}H)_3(\eta^5\text{-}C_5H_5)(CO)_9$ AND $CoOs_3(\mu\text{-}H)_4(\eta^5\text{-}C_5H_5)(CO)_9$

$$Os_3(\mu\text{-}H)_2(CO)_{10} + Co(\eta^5\text{-}C_5H_5)(CO)_2 \xrightarrow{H_2}$$
$$CoOs_3(\mu\text{-}H)_3(\eta^5\text{-}C_5H_5)(CO)_9 + CoOs_3(\mu\text{-}H)_4(\eta^5\text{-}C_5H_5)(CO)_9 \quad (2)$$

This reaction is carried out in a 50-mL, three-necked, round-bottom flask containing a magnetic stirrer, a gas inlet placed in one neck is tube positioned below the surface of the reaction solution, and a reflux condenser placed on a second neck is connected to a mercury–oil bubbler. Prepurified H_2 is bubbled through a toluene solution (20 mL) of 150 mg of $Os_3(\mu\text{-}H)_2(CO)_{10}$ (0.176 mmol). The system is flushed with H_2 for 30 min and 208 mg of $Co(\eta^5\text{-}C_5H_5)(CO)_2$ (1.156 mmol) is then added by means of a syringe through the third neck of the flask under the stream of issuing H_2. The third neck is then capped and with H_2 flowing through the solution, the solution is heated up to 90° and stirred. The time required for the reaction in the presence of H_2 depends on the flow rate of hydrogen. To make sure the reaction is complete, the reaction mixture is checked with TLC from time to time. For our particular parameters (H_2 flow rate of 50 mL/min at 298 K and 1 atm), 60 hr of reaction time is required. The solution is then cooled to room temperature. Then toluene and unconsumed $Co(\eta^5\text{-}C_5H_5)(CO)_2$ are pumped away under vacuum[13] to leave a dark brown residue that is dissolved in a minimum amount of CH_2Cl_2. Elution of this residue on a 2-mm preparative SiO_2 TLC plate with 1:4 toluene–hexane gives three bands, these being, in descending order of R_f value, purple $CoOs_3(\mu\text{-}H)_3(\eta^5\text{-}C_5H_5)(CO)_9$ (54.2 mg, 32%, $R_f = 0.83$), dark green $CoOs_3(\mu\text{-}H)_4(\eta^5\text{-}C_5H_5)(CO)_9$ (58.9 mg, 35%, $R_f = 0.64$), and green $CoOs_3(\mu\text{-}H)_2(\eta^5\text{-}C_5H_5)(\mu\text{-}CO)(CO)_9$ (2 mg, 1%, $R_f = 0.48$).

Properties of $CoOs_3(\mu\text{-}H)_3(\mu^5\text{-}C_5H_5)(CO)_9$ *and* $CoOs_3(\mu\text{-}H)_4(\eta^5\text{-}C_5H_5)(CO)_9$

$CoOs_3(\mu\text{-}H)_3(\eta^5\text{-}C_5H_5)(CO)_9$ and $CoOs_3(\mu\text{-}H)_4(\eta^5\text{-}C_5H_5)(CO)_9$ are obtained as purple and dark green solids, respectively. While stable in the solid state, CH_2Cl_2 solutions of these two compounds decompose slowly when exposed to air. Both are very soluble in most organic solvents such as hexane, benzene, and methylene chloride.

$CoOs_3(\mu\text{-}H)_3(\eta^5\text{-}C_5H_5)(CO)_9$ is paramagnetic. It contains one unpaired electron as determined by the Evans method.[14] The ESR signal is too broad (298–4K) to calculate a g value. The structure has been determined by X-ray diffraction[4,6,15] and spectral data are IR spectrum (hexane, ν_{CO}): 2082 (w), 2060 (s), 2008 (vs), 1990 (m), 1955 (vw) cm^{-1}. Mass spectrum: parent ion at m/e 954.8369, consistent with $^1H_8$ $^{12}C_{14}$ $^{16}O_9$ ^{59}Co $^{192}Os_3$ (calcd. value is 954.8336).

Anal. Calcd. 17.71 C, 0.85 H. Found: 17.50 C, 0.82 H.

The conditions chosen for reaction (2) permit significant yields of $CoOs_3(\mu\text{-}H)_3(\eta^5\text{-}C_5H_5)(CO)_9$. This compound, however, will react with H_2 to convert to $CoOs_3(\mu\text{-}H)_4(\eta^5\text{-}C_5H_5)(CO)_9$.[4] Spectral data of $CoOs_3(\mu\text{-}H)_4(\eta^5\text{-}C_5H_5)(CO)_9$ are IR spectrum (hexane, ν_{CO}): 2082 (m), 2060 (s), 2050 (s), 2019 (s), 1995 (s,sh), 1992 (s), 1977 (m), 1952 (vw) cm^{-1}. Mass spectrum: parent ion at m/e 955.8387, consistant with $^1H_9$ $^{12}C_{14}$ $^{16}O_9$ ^{59}Co $^{192}Os_3$ (calcd. value is 955.8415). ^1H NMR spectrum (CD_2Cl_2, $-60°$): $\delta = -5.23$ (s, 5H), -18.26 (s, 2H), 20.27 (s, 2H).

Anal. Calcd. 17.71 C, 0.95 H. Found: 17.49 C, .99 H.

References

1. J. S. Plotkin, D. G. Alaway, C. R. Weisenberger, and S. G. Shore, *J. Am. Chem. Soc.*, **102,** 6158 (1980).
2. M. R. Churchill, C. Bueno, W.-L. Hsu, J. S. Plotkin, and S. G. Shore, *Inorg. Chem.*, **21,** 1958 (1982).
3. L.-Y. Hsu, W.-L. Hsu, D.-Y. Jan, A. G. Marshall, and S. G. Shore, *Organometallics*, **3,** 591 (1984).
4. S. G. Shore, W.-L. Hsu, C. R. Weisenberger, M. L. Caste, M. Churchill, and C. Bueno, *Organometallics*, **1,** 1405 (1982).
5. S. G. Shore, W.-L. Hsu, M. R. Churchill, and C. Bueno, *J. Am. Chem. Soc.*, **105,** 655 (1983).
6. M. R. Churchill, C. Bueno, S. Kennedy, J. C. Bricker, J. S. Plotkin, and S. G. Shore, *Inorg. Chem.*, **21,** 627 (1982).
7. E. W. Burkhardt, and G. L. Geoffroy, *J. Organometal. Chem.*, **198,** 179 (1980).
8. M. Castiglioni, E. Sappa, M. Valle, M. Lanfranchi, and A. Tiripicchio, *J. Organometal. Chem.*, **241,** 99 (1983).
9. G. Lavigne, F. Papageorgiou, C. Bergounhou, and J. J. Bonnet, *Inorg. Chem.*, **22,** 2485 (1983).
10. L.-Y. Hsu, W.-L. Hsu, D.-Y. Jan, and S. G. Shore, *Organometallics*, **5,** 1041 (1986).
11. S. A. R. Knox, J. W. Koepke, M. A. Andrews, and H. D. Kaesz, *J. Am. Chem. Soc.*, **97,** 3942 (1975).

12. It is important that the reaction time not be drastically exceeded. $CoOs_3(\mu\text{-}H)_2(\eta^5\text{-}C_5H_5)(\mu\text{-}CO)(CO)_9$ will decompose in a toluene solution at 90° over an extended period of time.

13. $Co(\eta^5\text{-}C_5H_5)Co(CO)_2$ has to be removed completely from reaction mixture before the residue being eluted on TLC to have maximum yield of products.

14. J. Loliger, and R. Scheffold, *J. Chem. Educ.*, **49**, 646 (1972).

15. M. R. Churchill, and C. Bueno, *Inorg. Chem.*, **22**, 1510 (1983).

INDEX OF CONTRIBUTORS

SUBJECT INDEX

Names used in this Subject Index for Volumes 21–25 are based upon IUPAC *Nomenclature of Inorganic Chemistry*, Second Edition (1970), Butterworths, London; IUPAC *Nomenclature of Organic Chemistry*, Sections A, B, C, D, E, F, and H (1979), Pergamon Press, Oxford, U.K.; and the Chemical Abstracts Service *Chemical Substance Name Selection Manual* (1978), Columbus, Ohio. For compounds whose nomenclature is not adequately treated in the above references, American Chemical Society journal editorial practices are followed as applicable.

Inverted forms of the chemical names (parent index headings) are used for most entries in the alphabetically ordered index. Organic names are listed at the "parent" based on Rule C-10, *Nomenclature of Organic Chemistry*, 1979 Edition. Coordination compounds, salts and ions are listed once at each metal or central atom "parent" index heading. Simple salts and binary compounds are entered in the usual uninverted way, e.g., *Sulfur oxide* (S_8O), *Uranium(IV) chloride* (UCl_4).

All ligands receive a separate subject entry, e.g., *2,4-Pentanedione*, iron complex. The headings *Ammines, Carbonyl complexes, Hydride complexes,* and *Nitrosyl complexes* are used for the NH_3, CO, H, and NO ligands.

Borane, carboxy-compd. with trimethylamine, 25:81
——, cyano compd. with trimethylamine (1:1), 25:80
——, dichlorophenyl-, 22:207
——, diethylhydroxy-, 22:193
——, diethylmethoxy-, 22:190
——, (dimethylamino)diethyl-, 22:209
——, [(2,2-dimethylpropanoyl)oxy]diethyl-, 22:185
——, (ethylcarbamoyl)-trimethylamine (1:1), 25:83
——, ethyldihydroxy, *see* Boronic acid, ethyl-, 24:83
——, (methoxycarbonyl)-, compd. with trimethylamine, 25:84
——, (pivaloyloxy)diethyl, *see* Borane, [(2,2-dimethylpropanoyl)oxy]diethyl-, 22:185
[^{10}B]Borane, dibromomethyl-, 22:223
Borate(1−), cyanotri[(^{2}H)hydro]-, sodium, 21:167
——, (cyclooctane-1,5-diyl)dihydro-:
lithium, 22:199
potassium, 22:200
sodium, 22:200
——, dodecahydro-7,8-dicarba-*nido*-undecapotassium, 22:231
——, dodecahydro-6-thia-*arachno*-decacesium, 22:227
——, hydrotris(pyrazolato)-, copper complex, 21:108
——, tetrafluoro-:
μ-carbonyl-μ-ethylidyne-bis[carbonyl(η^{5}-cyclopentadienyl)ruthenium](1+), 25:184
dicarbonyl(η^{5}-cyclopentadienyl)(η^{5}-2-methyl-1-propenyl)iron(1+), 24:166
pentaammine(pyrazine)ruthenium(II) (2:1), 24:259
tetrabutylammonium, 24:139
4,4′,5,5′-tetramethyl-2,2′-bi-1,3-diselenolylidene, radical ion(1+) (1:2), 24:139
——, tetrakis(pyrazolato)-, copper complex, 21:110
——, tetraphenyl-,:
tetrakis(1-isocyanobutane)bis[methylenebis(diphenylphosphine)]dirhodium(I), 21:49

tetrakis(1-isocyanobutane)rhodium(I), 21:50
——, tris(3,5-dimethylpyrazolato)hydro-:
boron-copper complexes, 21:109
molybdenum complexes, 23:4–9
Borate(2−), tris[μ-[(1,2-cyclohexanedione dioximato)-*O*:*O′*]diphenyldi-:
iron complex, 21:112
Borate(III), tetrafluoro-, tetrafluoroammonium (1:1), 24:42
Borinic acid, diethyl, *see* Borane, diethylhydroxy-, 22:193
——, diethyl, methyl ester, *see* Borane, diethylmethoxy-, 22:190
Bornan-2-one, 3-*endo*-bromo-3-*oxo*-nitro-, (1*R*)-, 25:134
——, 3-*aci*-nitro-, (1*R*)-, sodium salt, 25:133
Boron, bis-μ-(2,2-dimethylpropanoato-*O*,*O′*)-diethyl-μ-oxo-di-, 22:196
——, tris[pentafluorooxotellurate(VI)]-, 24:35
[^{10}B] Boron bromide (^{10}BBr$_{3}$), 22:219
Boron compounds, labeling of, with boron-10, 22:218
Boronic acid, ethyl-, 24:83
Boroxin, triethyl-, mixture with tetraethyldiboroxane, (1:3), 24:85
Bromides, of rare earths and alkali metals, 22:1, 10
Bromoimidosulfurous difluoride, 24:20
1-Butanamine, intercalate with hydrogen pentaoxoniobatetitanate(1−), 22:89
Butane:
cobalt, iridium, and rhodium complexes, 22:171, 173, 174
palladium complex, 22:167, 168, 169, 170
——, isocyano-, rhodium complex, 21:49
1-Butanol, [(*N*,*N*-diethylcarbamoyl)methyl]phosphonate ester, 24:101
2-Butanol, 3-oxo-, dimethylcarbamodiselenoate ester, 24:132
Butyl, tin deriv., 25:112, 114
tert-Butyl alcohol, perfluoro-, hypochlorite ester, 24:61
tert-Butylamine, *N*-(trimethylsilyl)-, 25:8

Cadmate(II), tetrakis(benzenethiolato)-, bis-(tetraphenylphosphonium), 21:26

FORMULA INDEX

The Formula Index, as well as the Subject Index, is a Cumulative Index for Volumes 21–25. The Index is organized to allow the most efficient location of specific compounds and groups of compounds related by central metal ion or ligand grouping.

The formulas entered in the Formula Index are for the total composition of the entered compound, e.g., F_6NaU for sodium hexafluorouranate(V). The formulas consist solely of atomic symbols (abbreviations for atomic groupings are not used) and arranged in alphabetical order with carbon and hydrogen always given last, e.g., $Br_3CoN_4C_4H_{16}$. To enhance the utility of the Formula Index, all formulas are permuted on the symbols for all metal atoms, e.g., $FeO_{13}Ru_3C_{13}H_{13}$ is also listed at $Ru_3FeO_{13}C_{13}H_{13}$. Ligand groupings are also listed separately in the same order, e.g., $N_2C_2H_8$, 1,2-Ethanediamine, cobalt complexes. Thus individual compounds are found at their total formula in the alphabetical listing; compounds of any metal may be scanned at the alphabetical position of the metal symbol; and compounds of a specific ligand are listed at the formula of the ligand, e.g., NC for Cyano complexes.

Water of hydration, when so identified, is not added into the formulas of the reported compounds, e.g., $Cl_{0.30}N_4PtRb_2C_4 \cdot 3H_2O$.

$AgAsF_6$, Arsenate, hexafluoro-, silver, 24:74

$AgAsF_6S_{16}$, Silver(1+), bis(*cyclo*-octasulfur)-, hexafluoroarsenate, 24:74

$AgCoN_4O_8C_4H_8$, Cobaltate(III), bis-(glycinato)dinitro-, *cis*-(No$_2$), trans(N)-, silver(I), 23:92

AgF_3O_3SC, Silver trifluoromethanesulfonate, reactions, 24:247

Ag_3NO_3S, Silver(1+), μ_3-thio-tri-, nitrate, 24:234

$Ag_8O_{16}W_4$, Silver tungstate, 22:76

$AlBrSi_2C_8H_{22}$, Aluminum, bromobis-[(trimethylsilyl)methyl]-, 24:94

AlH_4LaNi_4, Aluminum lanthanum nickel hydride, 22:96

$AlNaO_4Si \cdot 2.25H_2O$, Sodium aluminum silicate, 22:61

———, Zeolite A, 22:63

$AlSi_3C_{12}H_{33}$, Aluminum, tris[(trimethylsilyl)methyl]-, 24:92

$Al_2Na_2O_{14}Si \cdot XH_2O$, Sodium aluminum silicate hydrate, 22:64

———, Zeolite Y, 22:64

$Al_{2.6}N_{3.6}Na_{2.4}O_{207}Si_{100}C_{43}H_{100}$, Sodium tetrapropylammonium aluminum silicate, 22:67

———, ZSM-5, 22:67

$Al_4K_2NNaO_{3.6}Si_{14}C_4H_{12} \cdot 7H_2O$, Offretite, tetramethylammonium substituted, 22:65

———, Potassium sodium tetramethylammonium aluminum silicate hydrate, 22:65

$AsAgF_6$, Arsenate, hexafluoro-, silver, 24:74

$AsAgF_6S_{16}$, Arsenate, hexafluoro-, bis-(*cyclo*-octasulfur)silver(1+), 24:74

$AsBr_3F_6S$, Arsenate, hexafluoro-, tribromosulfur(IV), 24:76

AsC_2H_7, Arsine, dimethyl-molybdenum complex, 25:169

$AsC_{18}H_{15}$, Arsine, triphenyl-chromium complexes, 23:38

$AsC_{24}H_{10}$, Arsonium, tetraphenyl-$1\lambda^4,3\lambda^4,5\lambda^4,7$-tetrathia-2,4,6,8,9-pentaazabicyclo[3.3.1]nona-1(8),2,3,5-tetraenide, 25:31

KCoN$_2$O$_8$C$_{11}$H$_{14}$, Cobaltate(III), [[R(−)]-N,N'-(1-methyl-1,2-ethanediyl)bis[N-(carboxymethyl)glycinato](4−)]-, [Δ-(+)]-, 23:101

KCoN$_2$O$_8$C$_{14}$H$_{18}$, Cobaltate(III), [[R,R-(−)]-N,N-1,2-cyclohexanediyl-bis(carboxymethyl)glycinato](4−)]-, [Δ-(+)]-, potassium, 23:97

KCoN$_4$O$_7$CH$_6$ · 0.5H$_2$O, Cobaltate(III), diamine(carbonato)dinitro-, *cis*,*cis*-, potassium, hemihydrate, 23:70

KCoN$_4$O$_8$C$_2$H$_6$ · 0.5H$_2$O, Cobaltate(III), diamminedinitro(oxalato)-, *cis*,*cis*-, potassium, hemihydrate, 23:71

KCoO$_2$, Potassium cobalt oxide, 22:58

KCl$_7$Dy$_2$, Potassium dysprosium chloride, 22:2

KCrO$_2$, Potassium chromium oxide, 22:59

KCr$_2$HO$_{10}$C$_{10}$, Chromate, μ-hydrido-bis[pentacarbonyl-, potassium, 23:27

KF$_6$U, Uranate(V), hexafluoro-, potassium, 21:166

KHO$_{10}$W$_2$C$_{10}$, Tungstate, μ-hydrido-bis[pentacarbonyl-, potassium, 23:27

KN$_2$O$_6$PC$_{14}$H$_{24}$, Potassium(1+), (1,4,7,10,13,16-hexaoxacyclooctade-cane)-dicyanophosphide(1−), 25:126

KNbO$_5$Ti, Potassium, pentaoxoniobateti-tanate(1−), 22:89

KSnC$_{12}$H$_{27}$, Stannate(1−), tributyl-potassium, 25:112

KSnC$_{18}$H$_{15}$, Stannate(1−), triphenyl-potassium, 25:111

K$_{0.5}$CoO$_2$, Potassium cobalt oxide, 22:57

K$_{0.5}$CrO$_2$, Potassium chromium oxide, 22:59

K$_{0.67}$CoO$_2$, Potassium cobalt oxide, 22:57

K$_{0.6}$CrO$_2$, Potassium chromium oxide, 22:59

K$_{0.7}$CrO$_2$, Potassium chromium oxide, 22:59

K$_{0.77}$CrO$_2$, Potassium chromium oxide, 22:59

K$_2$Al$_4$NNaO$_{36}$Si$_{14}$C$_4$H$_{12}$ · 7H$_2$O, Potassium sodium tetramethylammonium aluminum silicate hydrate, 22:65

K$_2$Bi$_4$N$_2$O$_{12}$C$_{36}$H$_{72}$, Potassium, (4,7,13,16,21,24-hexaoxa-1,10-diazabi-cyclo[8.8.8]hexacosane)-, tetrabismu-thide, 22:151

K$_2$F$_{0.60}$N$_4$PtC$_4$H$_{0.30}$ · 3H$_2$O, Platinate, tet-racyano-, potassium (hydrogen difluo-ride) (1:2:0.30), trihydrate, 21:147

K$_2$F$_5$Mn · H$_2$O, Manganate(III), penta-fluoro-, dipotassium, monohydrate, 24:51

K$_2$F$_5$MoO, Molybdate(V), pentafluorooxo-, dipotassium, 21:170

K$_2$I$_4$Pt · 2H$_2$O, Platinate(II), tetraiodo-dipotassium, dihydrate, 25:98

K$_2$Sb$_2$O$_{12}$C$_8$H$_4$, Antimonate(2−), bis[taratrato(4−)]di-, potassium as resolving agent, 23:76–81

K$_3$CrO$_{12}$C$_6$ · H$_2$O and 2H$_2$O, Chro-mate(III), tris(oxalato)-tripotassium, (+)-, dihydrate and (−)-, monohydrate, isolation of, 25:141

K$_4$H$_8$O$_{20}$P$_8$Pt, Platinate(II), tetrakis-[dihydrogen diphosphito(2−)]-, tetra-potassium, 24:211

K$_{10}$O$_{78}$P$_2$ThW$_{22}$, Thorate(IV), bis(undeca-tungstophosphato)-, decapotassium, 23:189

K$_{10}$O$_{78}$P$_2$UW$_{22}$, Uranate(IV), bis(undeca-tungstophosphato)-, decapotassium, 23:186

K$_{14}$B$_2$O$_{78}$ThW$_{22}$, Thorate(IV), bis(undeca-tungstoborato)-, tetradecapotassium, 23:189

K$_{16}$O$_{122}$P$_4$ThW$_{34}$, Thorate(IV), bis(heptade-catungstodiphosphato)-, hexadecapo-tassium, 23:190

K$_{16}$O$_{122}$P$_4$UW$_{34}$, Uranate(IV), bis(hepta-decatungstodiphosphato)-, hexadeca-potassium, 23:188

LaAlH$_4$Ni$_4$, Aluminum lanthanum nickel hydride, 22:96

LaCl$_3$, Lanthanum chloride, 22:39

LaF$_{18}$N$_6$O$_6$P$_{12}$C$_{72}$H$_{72}$, Lanthanium(III), hex-akis(diphenylphosphinic amide)-, tris(hexafluorophosphate), 23:180

LaI$_2$, Lanthanum iodide, 22:36

LaI$_3$, Lanthanum iodide, 22:31

LaN$_3$O$_{13}$C$_8$H$_{16}$, Lanthanium(III), trini-trato(1,4,7,10-tetraoxacyclododecane)-, 23:151

LaN$_3$O$_{14}$C$_{10}$H$_{20}$, Lanthanium(III), trini-trato(1,4,7,10,13-pentaoxacyclopenta-decane)-, 23:151

LaN$_3$O$_{15}$C$_{12}$H$_{24}$, Lanthanium(III), (1,4,7,10,13,16-hexaoxacyclooctade-cane)trinitro-, 23:153